GW00597747

Elementary Engineering Systems

B. F. Gray

Head of Department of Electrical Engineering and Physics, Hatfield Polytechnic

LONGMAN

LONGMAN GROUP LIMITED
London
Associated companies, branches and representatives
throughout the world

First published 1974

Library of Congress Catalog Card Number: 73-86519
ISBN 0582   41067.3   cased
            41068.1   paper

Set in Monotype Garamond
and printed in Great Britain
by Adlard & Son Limited
Bartholomew Press, Dorking

Preface

This book was originally written for the Systems section of the new Ordinary National Diploma in Technology (Engineering), realising that no book existed on the market specifically for students on this course. It therefore follows the laid down syllabus very closely, beginning with the elements of simple electrical, mechanical and fluid systems and building up to their interconnection, dynamic behaviour and control. There are, of course, a number of books on the subject of Systems already available but few if any written at a level comprehensible to the 18 or 19 year old whose mathematical knowledge is below 'A' level.

Hopefully this book will present the subject in a way which the average OND student will find relatively easy to follow.

Since completing the book it is apparent that students at a higher level might find this book a helpful introductory text to the subject of Engineering Systems.

It would, of course, be wrong not to record at this point the help (largely through discussion and debate) of my colleagues, the expert deciphering of my handwriting and the typing of the manuscript by Mrs Marion Dance and the long suffering of my wife and family while I withdrew to the study to write the book. I would also like to thank the Post Office, the British Aircraft Corporation and Electronic Associates Ltd who allowed me to reproduce the photographs included in the text, and finally the Belfast College of Technology who unknowingly supplied me with ideas for certain unworked examples.

Harpenden, July 1973

Acknowledgements

We are grateful to the following for permission to reproduce photographs:

Plates 1 and 6: The Post Office. Plates 2 and 3: The British Aircraft Corporation. Plates 4 and 5: E.A.L. Fig 8.1 is reproduced from *Value Engineering* by L. W. Crum; Longman 1971.

Contents

Chapter 1
Systems

1.1 DEFINITION AND EXAMPLES

The word 'system' is frequently used in day to day conversation, for example a transport system, an educational system, a bidding system used in contract bridge—to quote just a few usages. It is therefore necessary to say what we mean when we use the term.

A possible definition of a system is as follows: a system is a collection of parts, matter, components or data which are inter-related or connected and often interact within specified but sometimes arbitrary boundary conditions.

This is a rather general description of a system. We are going to concern ourselves with engineering systems but not necessarily those involving pieces of equipment or 'hardware'. Engineering systems can involve organisational aspects – for example the planning of a production line in a factory or the consideration of the problems involved in the production of a railway timetable.

Systems have inputs and outputs and they often respond in a more or less predictable manner. The simplest systems have one input and one output and the response to a given input can be completely defined. More complex systems may have a number of inputs and the behaviour may be less predictable. We may represent a system as a 'black box', thus providing a boundary, and show the input as a line with an arrow going into the box. The output is shown in a similar manner (see Fig. 1.1) but with the arrow coming out.

Fig. 1.1

If the input is θ_{in} (θ can be a force, voltage, flow of liquid or merely a mathematical expression) then θ_{out} is related to θ_{in} by the behaviour or characteristics of the system. $\theta_{out} = G\theta_{in}$ where G describes the effect the system has on θ_{in}. G is sometimes called the 'transfer function' of the system.

An example might be a loaded gun. The input is the pressure on the trigger. This results in a mechanical impact on the cartridge, a violent chemical reaction, a rapid expansion of gases in the barrel and the expulsion of the bullet. The behaviour is very predictable!

Perhaps a good example of a more complex engineering system is the motor car. It consists of a metal framework forming the chassis and body – the physical boundary of the system – and within this framework are a number of smaller systems or subsystems – the engine, gearbox, clutch and transmission – forming a complicated mechanical system; the ignition circuit, battery, generator and lighting – forming an electrical system; the dashboard, steering column, gear lever and foot pedals – forming part of a control system; and finally there is a suspension system connecting all this to the road wheels.

These subsystems all act in a largely predictable manner. They are interconnected and all contribute to the smooth running of the car but not all are vital to the operation of the system as a whole. Some are more important than others. For example a fault in the lighting system will not hamper the movement of the car, a faulty suspension will not render the car useless (although uncomfortable to ride in). On the other hand a fault in the ignition circuit or a completely worn clutch will immobilise the car until the fault is repaired.

Another example of a complex engineering system is a spacecraft. Such a vehicle contains a vast number of systems – guidance, propulsion, communications, life support, etc. A complicated system can always be broken down to smaller and less complex organisations, components, or elements which can then be easier to understand and analyse.

Man himself is a very complicated biological system.

Some idea of the vast field covered by the term 'systems' is illustrated in Fig. 1.2. Engineering systems cover a selection of these indicated areas.

1.2 SOME PHILOSOPHICAL VIEWS

We usually desire to have systems under some form of control but occasionally the system proves too complex to do this. The economic situation within a country may be controllable but tends to be affected by world events or world economics which are outside the control of the country. We need to know if a system is stable or unstable – whether it is a moving or changing system in order that we can at least attempt to control it to prevent possibly catastrophic results.

We need certain systems in order to have some control over our environment (e.g. a heating system), to be able to travel (a transport system), to be able to talk to each other at great distances (a communication system). Some systems serving a useful purpose also pose

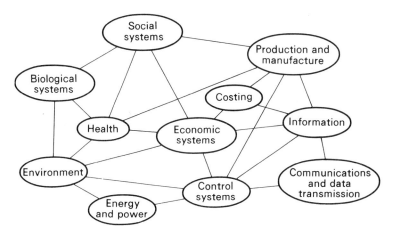

Fig. 1.2

problems (the useful car produces exhaust fumes which pollute the air which we breathe); other systems are employed in a manner which raises doubts in some people's minds concerning the purpose or usefulness of the system – advertising on our TV networks for example may be regarded by some as an abuse of a communication system.

We have to employ man-made systems in the world in which we live, for without them we would tend to revert to our primitive origin and be at the mercy of our environment. It is paradoxical that technology may be guilty of affecting the environment and we may therefore be at the mercy of a new and potentially hostile environment of our own making.

There is however a 'happy medium' in which systems work for the overall benefit of man and man is not the slave or victim of the systems of his making. It is as well to attempt to understand systems, to know how they are built up, to be able to make measurements on them, to know how they function, in order to be able to control them and live in harmony with them.

1.3 THE USE OF MODELS

In order to understand a system it is often a useful exercise to build a 'model' which behaves in the same way as (or at least in a similar way to) the system being examined. The model is frequently far cheaper than the system it represents and is invariably more convenient to use for experimental purposes. It can be a mathematical model which expresses the system as a number of variable quantities in an equation. By using well known methods of analysis – or by changing the mathematical model to a form acceptable to a computer – we can see the behaviour of the model to an applied stimulus, or test whether the system is stable, or change the parameters of the system to bring about a desired result. We can then apply the results of our observations on the model to the system itself. If (a big 'if'!) the model is an accurate replica of the system then the system will respond in precisely the same way as the model. One cannot be absolutely sure that this will happen because of the unknown inaccuracies in the building of the model.

The model can be a scaled down version of the system. We need to know how the Concorde (an aero system) will behave at high speed. We make a scaled down model and place it in a wind tunnel and observe its behaviour. Although we are reasonably sure that the full-scale Concorde will behave in a similar manner we have to test it at high speed in the air under actual operating conditions before we can be sure. A subtle but vital point may be missing from the model.

Sometimes we may be dealing with abstract ideas and the model is not a true presentation of the facts but some means of enabling us to understand the subject matter better. The model of an atom is such an example.

1.4 ENGINEERING SYSTEMS

It is convenient to approach the study of engineering systems by firstly considering specific components or elements, deciding how they behave and then linking them together to form first simple and then complex systems. It must be appreciated that subsystems may interact so that the behaviour of a complex system is not necessarily the simple sum of the behaviours of individual subsystems. This approach could be described as one of synthesis.

An alternative method of study would be to consider an existing system where we can determine the behaviour of the complete system. We can then break down the system to smaller units in order to examine the subsystem behaviour and finally the components themselves. This is an analytical approach. In this book we shall adopt the former.

Engineering components can be grouped into the following categories, mechanical, electrical, fluid, thermal, etc., although only in the simplest cases can they be regarded as exclusively of one type. A spring is a mechanical component or element, a coil is an electrical component or element. Put them together to form a relay and we then have an electromechanical component. Link a number of relays together and we may then have a telephone exchange – part of a communication system.

We are starting by considering the simplest of cases, first mechanical systems.

1.5 ELEMENTS OF SYSTEMS

MECHANICAL SYSTEMS

A mechanical system has three basic elements: mass m which possesses inertia, springs which possess stiffness k and dampers which possess resistance a (Fig. 1.3). When these elements are interconnected we then have a mechanical system which behaves in a predictable way. In the simplest case it is possible to describe each element by a mathematical relationship – a mathematical model in fact.

Let us consider the effect of mass first of all.

In an idealised case we say that the force F (in newtons) required to accelerate a mass m (in kilograms) is given by

$$F = mf \qquad \text{where } f \text{ is the acceleration of the mass (in metres per second squared)}$$

f can also be defined as the change of velocity u with time t, i.e.

$$f = \frac{u}{t} \quad \text{or better} \quad \frac{du}{dt}$$

u can be defined as the change of distance s with time, i.e.

$$u = \frac{s}{t} \quad \text{or better} \quad \frac{ds}{dt}$$

so that

$$f = \frac{du}{dt} = \frac{d^2s}{dt^2}$$

Hence

$$F = m\frac{du}{dt} \quad \text{or} \quad m\frac{d^2s}{dt^2} \tag{1}$$

This is the force required to overcome the inertia of the system to cause it to move or change its velocity. In this equation F and s are variable quantities varying with time t, and mass m is a constant. (There are of course instances where m can vary with time – for example in rocket propulsion, but we will not consider such cases at this point.)

Let us now turn our attention to a simple helical spring.

A spring requires a force to extend it. Providing that we work well within the elastic limit the force F (newtons) to extend a spring an amount s (metres) is proportional to s. This is a statement of Hooke's law.
Hence

$$F = ks \qquad (2)$$

k is called the stiffness of the spring and is measured in newtons per metre. F and s are variable quantities, k is a constant.

The third element in our mechanical system is damping. Damping tends to slow down movement – rather like the effect of thick oil on the movement of a rod placed in it. One form of mechanical damping device is the 'dashpot' as illustrated in Fig. 1.3. The small hole in the

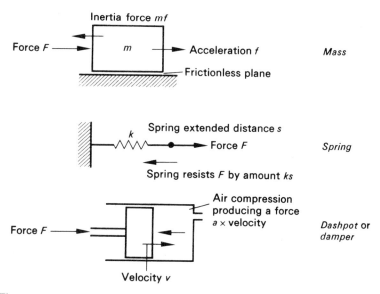

Fig. 1.3

cylinder head allows air to escape from the closed cylinder as we push the piston in. Similarly as we pull the piston out and tend to create a partial vacuum in the cylinder, air rushes in. A cushioning effect is produced.

In the ideal case the force required to push (or pull) the piston is directly dependent upon the speed at which we operate the piston. This is called **Viscous damping**.

Thus the force F (newtons)$= a \times$ velocity (metres per second)

$$F = a \frac{\mathrm{d}s}{\mathrm{d}t} \tag{3}$$

a is called the damping coefficient and is measured in newton seconds per metre. Again F and s are variables, varying with time t, and a is considered constant.

1.6 ENERGY CONSIDERATIONS

It should be noted that energy is stored in the mass when it is moving.

The kinetic energy (K.E.) is given by $\frac{1}{2}mu^2$ or $\frac{1}{2}m(\mathrm{d}s/\mathrm{d}t)^2$ joules. (*Note* $(\mathrm{d}s/\mathrm{d}t)^2$ is not the same as $\mathrm{d}^2s/\mathrm{d}t^2$.) The spring when extended also possesses energy and will return to its original position when released. It stores potential energy (P.E.). The energy stored by a spring is given by $\frac{1}{2}ks^2$ joules and this can be proved as follows:

> The force F to stretch a spring a distance s is given by $F = ks$ (newtons).
> Work or energy moving a distance $\mathrm{d}s$ is given by $F\,\mathrm{d}s$.
> Thus energy stored in the spring by extending it a distance $\mathrm{d}s$ is given by $ks\,\mathrm{d}s$ joules.
> In extending it a distance s metres the total energy is given by
>
> $$\int_0^s ks\,\mathrm{d}s = \tfrac{1}{2}ks^2 \text{ joules.}$$

The dashpot cannot store energy; it does not automatically return to its original position and therefore dissipates rather than stores energy. Work has to be done when we move the piston either in or out and results in a change of temperature of the air in the cylinder.

1.7 ELEMENTS OF ELECTRICAL ENGINEERING SYSTEMS

The three main elements in the electrical system which are analogous to the ones in the mechanical system are an inductor which produces a magnetic field and stores energy, a capacitor which produces an electric field and again stores energy and a resistor which dissipates energy (in the form of heat). The voltage v (volts) across an inductor depends upon the change in current i (amperes) through it.

Thus

$$v = L\frac{\mathrm{d}i}{\mathrm{d}t}, \quad \text{where } L \text{ is the inductance in henrys. But charge } q \text{ can be expressed as } q = it \text{ coulombs.}$$

Hence

$$i = \frac{q}{t} \quad \text{or} \quad \frac{\mathrm{d}q}{\mathrm{d}t}$$

Therefore

$$\frac{\mathrm{d}i}{\mathrm{d}t} = \frac{\mathrm{d}^2q}{\mathrm{d}t^2}$$

Hence

$$v = L\frac{\mathrm{d}^2q}{\mathrm{d}t^2} \tag{4}$$

This equation should be compared with (1). v is a variable – so is q, both changing with time – but L is assumed constant.

The voltage across a capacitor depends upon the charge it possesses. Thus

$$v = \frac{1}{C}q \tag{5}$$

where C is the capacitance in farads. *This equation is analogous with equation* (2) relating force and stiffness of a spring, force F is replaced by voltage v and distance s is replaced by charge q. k is replaced by $1/C$.

The voltage across a resistor is given by $v = iR$ where R is the resistance in ohms, but $i = \mathrm{d}q/\mathrm{d}t$.
Hence

$$v = R\frac{\mathrm{d}q}{\mathrm{d}t} \tag{6}$$

1.8 ENERGY CONSIDERATIONS

It should be noted that energy is stored in the inductor and is given by $\frac{1}{2}Li^2$ joules (or $\frac{1}{2}L(\mathrm{d}q/\mathrm{d}t)^2$). This can be shown as follows:

The voltage across an inductor caused by a change in current is $L(\mathrm{d}i/\mathrm{d}t)$ and if the current at an instant is i the power is given by the product voltage and current, i.e. $L(\mathrm{d}i/\mathrm{d}t) \times i$ watts.

During the interval dt the energy stored (product of power and time) is

$$L\frac{di}{dt} \times i \times dt = Li\, di \text{ joules}$$

Therefore the total energy stored while the current changes from o to i is

$$\int_0^i Li\, di = \tfrac{1}{2}Li^2 \text{ joules}$$

The energy stored in the capacitor is $\tfrac{1}{2}(q^2/C)$ joules. This can be shown as follows:

The voltage across a capacitor is given by q/C. If a change in charge dq/dt occurs current flows into or out of the capacitor either charging or discharging. Since $i = dq/dt$ it follows that the product of current and voltage (the power) is $(q/C)(dq/dt)$. During the interval dt the energy stored is

$$\frac{q}{C}\frac{dq}{dt}dt = \frac{q}{C}dq \text{ joules}$$

Therefore: the total energy stored while the charge changes from o to q is

$$\int_0^q \frac{1}{C}q\, dq = \frac{1}{2}\frac{q^2}{C} \text{ joules}$$

The resistor is a dissipative element and does not store energy.

One can now see a strong similarity between the mechanical elements of mass (inertia), stiffness and damping and the electrical elements inductance, capacitance and resistance (see Fig. 1.4):

mass m corresponds to inductance L;
stiffness k corresponds to the reciprocal of capacitance, $1/C$;
damping a corresponds to resistance R; also
distance s (a variable) corresponds to charge q (a variable).

(It is possible to derive an alternative table of analogous quantities in the electrical and mechanical systems where force and current are regarded as equivalent rather than force and voltage. It is not such an obvious analogy although it has certain advantages. For the present we will stick to the relationships derived.)

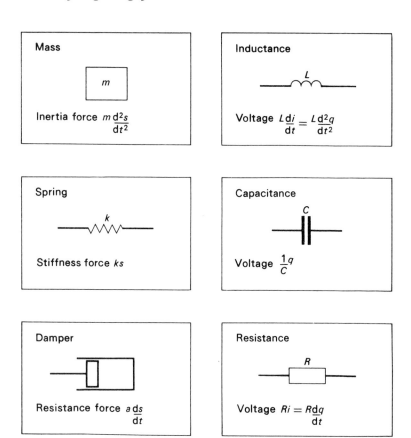

Fig. 1.4

1.9 ELEMENTS OF FLUID SYSTEMS

Energy may be stored in a tank by filling it with a liquid. The pressure on the bottom of the tank is dependent upon the volume or quantity Q of liquid and its density.

Thus: Pressure $P = \dfrac{Q \rho g}{\text{area } A}$ newtons per square metre (see Fig. 1.5)

Q = volume in cubic metres
A is in square metres, g = gravitational constant (9.81 m/s^2)
ρ the density is in kilograms per cubic metre.

Thus: $Q = \dfrac{AP}{\rho g}$

or: $Q = CP$ where C is the 'capacity' of the system, $A/\rho g$.

Hence

$$P = \frac{Q}{C} \tag{7}$$

(The units for this 'capacity' will be (metre)^5 per newton.)

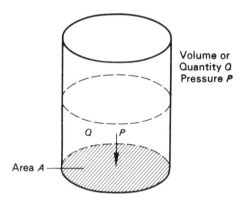

Fig. 1.5

This situation is similar to the electric case, P being the fluid pressure corresponding to electrical pressure (voltage) and Q being the quantity of water corresponding to charge q. The fluid system has a quantity 'inertance' which is similar to the inductance of the electrical system.

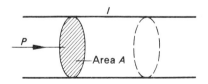

Fig. 1.6

Consider a length of pipe containing incompressible fluid, Fig. 1.6. If the pipe has uniform cross-sectional area A square metres and a length l of the pipe is considered then the mass of liquid over this length is given by $\rho l A$. Thus the force F to accelerate this mass is

$$F = \rho l A \frac{\mathrm{d}u}{\mathrm{d}t} \quad \text{or} \quad \frac{F}{A} = \rho l \frac{\mathrm{d}u}{\mathrm{d}t} \quad \text{where } u = \text{velocity}$$

$$\frac{F}{A} = \text{pressure } P \text{ hence } P = \rho l \frac{\mathrm{d}u}{\mathrm{d}t}$$

Now the velocity = distance l moved by a given volume of liquid in time $= l/t$ or $\mathrm{d}l/\mathrm{d}t$. But $Q = Al$.

Hence $\qquad l = \dfrac{Q}{A}$ and $\dfrac{\mathrm{d}l}{\mathrm{d}t} = \dfrac{1}{A}\dfrac{\mathrm{d}Q}{\mathrm{d}t}$ for a uniform cross-section A

$\qquad\qquad\qquad\qquad\qquad\qquad\qquad\qquad \mathrm{d}Q/\mathrm{d}t$ is the 'flow',

Therefore $\qquad \dfrac{\mathrm{d}u}{\mathrm{d}t} = \dfrac{1}{A}\dfrac{\mathrm{d}^2Q}{\mathrm{d}t^2}$

Hence:

$$P = \frac{\rho l}{A}\frac{\mathrm{d}^2Q}{\mathrm{d}t^2} = I\frac{\mathrm{d}^2Q}{\mathrm{d}t^2} \tag{8}$$

where I is the inertance of the system similar to the inductance in the electrical case. The units for I would be newton seconds squared per (metre)[5].

If a constriction is placed in a pipe a resistance is presented to the flow of liquid. The pressure needed to force the quantity Q through the constriction is approximately proportional to the velocity of liquid flow.

Since velocity u has already been shown to be proportional to $\mathrm{d}Q/\mathrm{d}t$ it follows that

$$u = R\frac{\mathrm{d}Q}{\mathrm{d}t} \tag{9}$$

where R is the 'resistance' offered to the fluid. The units for R would be newton seconds per (metre)[5].

1.10 ENERGY CONSIDERATION

The close analogy of the mechanical electrical and now fluid elements suggests the next step. From the previous work we may make an informed guess about the energy storage in a fluid system.

The energy stored in the fluid movement (K.E.)

$$\tfrac{1}{2}I\left(\frac{\mathrm{d}Q}{\mathrm{d}t}\right)^2 \text{ joules}$$

The energy stored in the vessel (P.E.)

$$\tfrac{1}{2}\frac{Q^2}{C} \text{ joules}$$

The resistance (or constriction) is a dissipative element.

We can now make a comparison of the three systems – mechanical, electrical and fluid. See Table 1.1.

Mechanical	Electrical	Fluid
Force F	Voltage V	Pressure P
Variables:	Variables:	Variables:
distance s	charge q	quantity Q
velocity $v = \dfrac{ds}{dt}$	current $i = \dfrac{dq}{dt}$	flow $\dfrac{dQ}{dt}$
Inertia force	Inductance voltage	Inertance pressure
$\dfrac{md^2s}{dt^2}$	$\dfrac{Ld^2q}{dt^2}$	$\dfrac{Id^2Q}{dt^2}$
Stiffness force	Capacitance voltage	Capacity pressure
ks	$\dfrac{1}{C}q$	$\dfrac{1}{C}Q$
Damping force	Resistance voltage	Resistive pressure
$a\dfrac{ds}{dt}$	$R\dfrac{dq}{dt}$	$R\dfrac{dQ}{dt}$

Table 1.1

1.11 ELEMENTS OF THERMAL SYSTEMS

It is possible to set up further analogies, for example in thermal systems, with the systems just discussed, the variable being temperature in this case. It will suffice to say at this stage that if the analogy exists then it is possible to set up an equivalent mechanical, electrical or fluid system which behaves similarly to a thermal system.

1.12 LINEARITY

The various elements considered so far have all been assumed constant. Mass m, stiffness k and damping a in the mechanical case, inductance L, capacitance C and resistance R in the electrical case, and inertance I, capacity C and resistance R in the fluid case.

The variables have been force and distance in the mechanical system, voltage and charge in the electrical system and pressure and quantity in the fluid case. We have seen how these variables all depend upon time t and it follows that t is an 'independent' quantity or variable on which the other variables depend. We are noting how they change or vary with time. Thus for example:

$$\frac{ds}{dt} \quad \text{(the speed) controls the damping force}$$

$$\frac{dq}{dt} \quad \text{(the current) controls the resistive voltage}$$

$$\frac{dQ}{dt} \quad \text{(the flow) controls the pressure drop across a restriction}$$

We can generalise here – let the fundamental variable be parabolic with time, i.e. $\propto t^2$ (Fig. 1.7). Then the graphs showing the derived

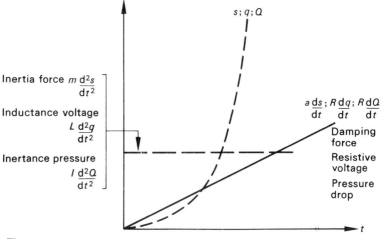

Inertia force $m \dfrac{d^2s}{dt^2}$

Inductance voltage $L \dfrac{d^2q}{dt^2}$

Inertance pressure $I \dfrac{d^2Q}{dt^2}$

$s; q; Q$

$a \dfrac{ds}{dt}; R \dfrac{dq}{dt}; R \dfrac{dQ}{dt}$

Damping force

Resistive voltage

Pressure drop

Fig. 1.7

forces, voltages and pressures are as shown if the elements are constant or 'linear'.

It is of course possible to have elements which do not remain constant, e.g. the current through a rectifier. Here v is not proportional to i and the resistance of the device varies. The resistive voltage across such an element would *not* increase linearly with time even if the current through it increased in a linear fashion. The waveform for example of current for a triangular voltage wave applied to such a non-linear element would look like that shown in Fig. 1.8.

The analysis for the behaviour of non-linear elements in a system poses many difficult problems – well beyond the reach of this book.

1. Post Office Tower, London

2. Intelstat IV Communications Satellite

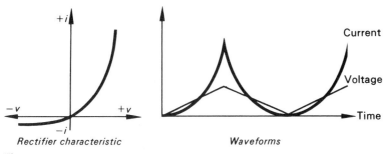

Rectifier characteristic   Waveforms

Fig. 1.8

We will deal only with idealised elements fully recognising the fact that in practice departures from the ideal will almost always occur to some extent and show up the limitations of any mathematical model which we are able to use at this stage.

One final comment. The elements mass, stiffness and the corresponding elements in the electrical and fluid cases are all non-dissipative, that is they can store energy, they do not dissipate it. Dissipative elements, damping and resistance, do not store energy.

1.13 SUMMARY

Mechanical systems

Elements or components	mass, stiffness, damping
Variables	force and distance (velocity)

Electrical systems

Elements or components	inductance, reciprocal of capacitance, resistance
Variables	voltage and charge (current)

Fluid systems

Elements or components	inertance, reciprocal of capacity, resistance
Variables	pressure and quantity (flow)

All variables are dependent upon time t. Velocity is the rate of change of distance, current is the rate of change of charge and flow is the rate of change of quantity.

Elements or components are considered constant.

Non-dissipative elements can store energy.

Dissipative elements (resistance and damping) absorb energy.

QUESTIONS

1.1 Identify the models in the following list indicating the object which they model.

(a) A walk on the Yorkshire moors.
(b) $y = mx + c$.
(c) A polytechnic.
(d) A photograph.
(e) A straight line.
(f) A transistor radio.
(g) Money.
(h) A family group.
(i) An Ordnance Survey map.
(j) A circuit diagram.
(k) A blueprint.
(l) H_2O.
(m) A car.
(n) Male.
(o) A prospectus.
(p) £1 note.
(q) Water.
(r) ○→.

1.2 Produce a simple model of the UK education system showing primary, secondary and tertiary education.
What are the inputs and outputs of the system?

1.3 What elements in a thermal system compare with mass, stiffness and damping in a mechanical system?

1.4 A spring of stiffness 500 N/m is arranged vertically and a mass of 4 kg is suspended from the free end. What is the stored energy in the spring?

1.5 An inductance of 5 H carries a current of 9 A. If this energy were transferred without loss to a capacitance of 15 μF what would be the voltage across the plates?

1.6 A cubic tank measures 5 m along each side and is placed on a horizontal plane. It is filled with water. Using the horizontal plane as reference what is the energy stored and the pressure on the base of the tank?

1.7 What are the units of capacity, resistance and inertance in the fluid system?

1.8 A voltage of 6 V is applied to a wire of resistance 5 Ω for 1 h. What energy is dissipated? What value of capacitance would be needed to store the same amount of energy from the same supply? Is this a practical value?

ANSWERS

1.4 1·54 J   1.5 5 196 V   1.6 6·13 MJ; 0·0491 MN/m²
1.8 25·9 kJ; 1 439 F

Chapter 2
Components

2.1 LINKED COMPONENTS

When elements or components are linked together they form a system and the behaviour of the system is governed by their interaction.

We can produce either mechanical, electrical or fluid systems by linking the respective elements. More complicated systems are produced by interaction of elements from more than one type of system – for example an electric motor driving a hoist may be regarded as an electromechanical system.

2.2 MECHANICAL SYSTEMS

Consider first a simple parallel mechanical system which links together a spring and a dashpot. The system consists of a bar B of negligible mass attached to a spring and a damper (Fig. 2.1). The other ends of the spring and damper are attached to a rigid wall. The spring is initially unextended.

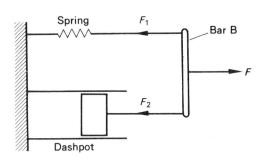

Fig. 2.1

A steady force F is suddenly applied as shown in the diagram. What will happen to the bar B? Fairly obviously it will move to the right but will it move suddenly?

Let us assume that the mass of the spring and the damper piston may be neglected and that the bar has moved a distance s metres from its original position in time t seconds.

The restraining forces of the spring and the dashpot, F_1 and F_2, oppose F so that

$$F = F_2 + F_1$$

Now $F_1 =$ stiffness force ks newtons

$F_2 =$ damping force $a(\mathrm{d}s/\mathrm{d}t)$ newtons

Hence

$$F = a\frac{\mathrm{d}s}{\mathrm{d}t} + ks \qquad (10)$$

This is a first-order differential equation and is a mathematical model of the system. It can be solved by standard mathematical methods and has the solution

$$s = \frac{F}{k}(1 - e^{-tk/a}) \qquad (11)$$

This result states that the distance moved by the bar depends upon time t, stiffness k and damping a. It is also directly dependent upon the initial force F. Ultimately $s = F/k$.

The solution can be shown to fit the original equation as follows:

If $\qquad s = \dfrac{F}{k}(1 - e^{-tk/a})$ then $\dfrac{\mathrm{d}s}{\mathrm{d}t} = \dfrac{F}{k} \times \dfrac{k}{a} e^{-tk/a}$

and $\qquad ks = F(1 - e^{-tk/a})$

$\therefore \qquad a\dfrac{\mathrm{d}s}{\mathrm{d}t} = F e^{-tk/a}$

$\therefore \qquad a\dfrac{\mathrm{d}s}{\mathrm{d}t} + ks = F e^{-tk/a} + F(1 - e^{-tk/a}) = F$

which is the original equation.

2.3 ELECTRICAL SYSTEMS

Consider the electrical system shown in Fig. 2.2. Here two electrical elements, a resistance and capacitance, are connected in series.

If the capacitance is initially uncharged and a sudden voltage E is applied by closing the switch how will the charge q on the capacitor build up?

The voltage v_C across the capacitor C at some time t seconds after closing the switch is $(1/C)\,q$. The voltage v_R across the resistor R is

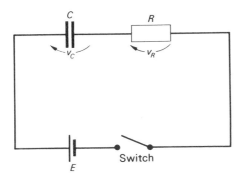

Fig. 2.2

Ri where i is the current flowing in the circuit. But $i = dq/dt$ hence $v_R = R(dq/dt)$.

These two voltages v_R and v_C must add up to the applied voltage E. Thus

$$E = R\frac{dq}{dt} + \frac{1}{C}q \qquad (12)$$

This equation is similar to (10). We have another first-order differential equation and we can obtain the solution by analogy with the previous example. E compares with F; C with $1/k$; R with a.

Therefore the solution is

$$q = CE(1 - e^{-t/CR}) \qquad (13)$$

The charge q depends upon the applied voltage and the capacitance but is time dependent. Ultimately $q = CE$.

2.4 FLUID SYSTEMS

Consider the fluid system in Fig. 2.3. The tank on the left supplies water at a constant pressure P_m (which is indicated by the height of liquid, H). If the valve is suddenly opened and the rate of flow of liquid is sufficiently small to neglect the inertance we can say that the pressure on the base of the right-hand tank (P_{tank})+pressure drop in the connecting tube (which offers a resistance P_{tube}) must equal the constant pressure P_m or $P_m = P_{tube} + P_{tank}$.
Hence

$$P_m = R\frac{dQ}{dt} + \frac{1}{C}Q \qquad (14)$$

another first-order differential equation.

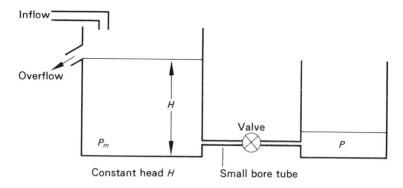

Fig. 2.3

The solution of this equation can be quoted by analogy

$$Q = CP_m(1 - e^{-t/CR}) \tag{15}$$

2.5 GENERALISATION OF THE RESULTS

The three systems are analogous and produce similar results and can be described by the same mathematical equation. In practice the three systems will behave slightly differently from the mathematical model because we have neglected mass in the first system, inductance in the second and inertance in the third. Nevertheless in the ideal case we need consider only one system and the conclusions we reach will apply equally well to the other two.

Since the electrical system is the easiest one to set up and make measurements on we will make this the one for discussion.

Firstly the growth of charge (or displacement s, or quantity of liquid, Q) follows an exponential curve illustrated in Fig. 2.4. It will be noticed that the final limit for q is CE when t approaches infinity. If one considers the rate of change of charge dq/dt (this is the current flowing) then we get

$$\frac{dq}{dt} = \frac{E}{R} e^{-t/CR}$$

when $t=0$ $dq/dt=E/R$ and hence the initial growth rate of the charge is determined by the applied voltage E and the resistance R. In the same way the initial growth rate of Q in the fluid case (the flow) is P_m/R and the initial rate of change of s (the velocity) is F/a in the mechanical system. Study one system and you have the lot!

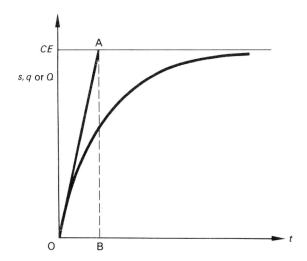

Fig. 2.4

The final values achieved in each case are respectively

$$q=CE \qquad Q=CP_m \qquad s=\frac{1}{k}F$$

which are also the initial disturbances or 'step inputs'.

It will be seen from the diagram that the initial slope of the exponential curve is given by AB/OB.

Thus $\dfrac{E}{R}$ (the slope at $t=0$)$=\dfrac{AB}{OB}=\dfrac{CE}{OB}$

Hence $\quad OB=\dfrac{CE}{E/R}=CR$

This quantity is called the 'time constant' of the circuit and indicates the time it would take for the charge to reach its final value if the initial growth were maintained. If C is in farads and R in ohms the time constant is in **seconds**.

The time constant for the fluid system is also CR seconds and for the mechanical system a/k seconds.

If we wish to 'slow down' a system such as the ones we have just dealt with we increase the resistance or damping, increase the capacitance but *decrease* the stiffness.

The general form of the solution to these equations may be derived as follows.

Taking for example the solution in the electrical case

$$q=CE(1-e^{-t/CR})$$

Now $\quad q = C v_C$ and letting $CR =$ time constant τ

$$C v_C = CE(1 - e^{-t/\tau})$$

or $\qquad v_C = E - E\, e^{-t/\tau}$

E is the step input, $v_C =$ value of the variable quantity under consideration so that in general

$$x = X - X\, e^{-t/\tau} \text{ following a step input } X$$

x is therefore composed of a steady value X which does not depend upon time and a variable quantity whose decay rate depends upon the time constant of the system.

3. Europe's Cosmic Ray Satellite

2.6 ENERGY CONSIDERATIONS

Energy is stored in the non-dissipative element in each case (the capacitor, the spring or the tank). Energy is dissipated in the dissipative element (the resistor, the dashpot or the constriction). The stored energy is all dissipated in the dissipative element but the rate of transfer of the energy depends upon the time constant of the system.

2.7 SECOND-ORDER SYSTEMS

Consider the mechanical system illustrated in Fig. 2.5. Here the damping dashpot has been replaced by a mass m kilograms. The system can be slightly rearranged (Fig. 2.6) to produce a rather improved design without changing the basic function of the system.

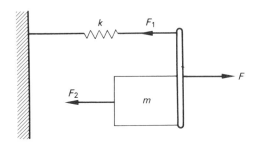

Fig. 2.5

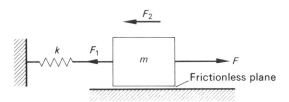

Fig. 2.6

A steady force F has been applied displacing the mass m a distance s from its original position. This stretches the spring and creates a restoring force. The force F is suddenly removed when the mass has been displaced a distance A from its original position. What happens?

F_1 due to the spring $= ks$

F_2 due to the inertia of the mass $= m \dfrac{\mathrm{d}^2 s}{\mathrm{d}t^2}$

These forces add up to the applied force F

Hence $F = m \dfrac{d^2 s}{dt^2} + ks$

and when F is removed at time $t = 0$

$$m \frac{d^2 s}{dt^2} + ks = 0 \qquad (16)$$

We may observe that this is an equation which involves the second derivative $d^2 s/dt^2$. This type of equation has been met with before in (for example) the swinging pendulum and describes simple harmonic motion (S.H.M.). The solution of such an equation is

$s = A \cos \omega t$ where A and ω are constants still to be determined.

A and ω can be found by substitution in the original equation as follows:

If $s = A \cos \omega t$; $\dfrac{ds}{dt} = -\omega A \sin \omega t$; $\dfrac{d^2 s}{dt^2} = -\omega^2 A \cos \omega t$

Substituting in $F = m \dfrac{d^2 s}{dt^2} + ks$

we have $F = -\omega^2 m A \cos \omega t + kA \cos \omega t$

when $F = 0$ (i.e. the restraining force is removed)

$$0 = -\omega^2 m A \cos \omega t + kA \cos \omega t$$

This equation is true for all values of t. If therefore we choose a particular value of t we can find the constants.
By letting $t = 0$, $\cos \omega t = 1$

$\therefore$ $0 = -\omega^2 m A + kA$

$0 = -\omega^2 m + k$ (since A cannot be zero)

$\therefore$ $\omega^2 = \dfrac{k}{m}$ $\omega = \sqrt{\dfrac{k}{m}}$

Now $s = A \cos \omega t$.
When $t = 0$, $s = A$ and this is obviously the initial displacement of the spring.
From $F = ks$ it follows that $s = A = F/k$, i.e. the initial restraining force divided by the stiffness. Therefore the full solution is

$$s = A \cos \omega t \quad \text{where} \quad A = \frac{F}{k} \quad \text{and} \quad \omega = \sqrt{\frac{k}{m}} \qquad (17)$$

This result shows that a spring/mass system just described oscillates with simple harmonic motion at an angular frequency of $\sqrt{k/m}$ radians per second or a cyclic frequency of $1/2\pi\sqrt{k/m}$ cycles per second or hertz with an amplitude F/k (the initial displacement) (Fig. 2.7).

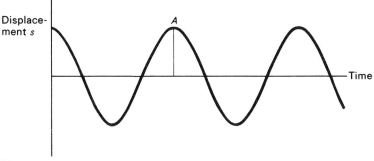

Fig. 2.7

In practice the oscillations would die away due to frictional forces not considered but in the ideal case such oscillations would continue indefinitely at a constant amplitude.

2.8 ELECTRICAL SECOND-ORDER SYSTEMS

A similar electrical system is shown in Fig. 2.8. Here the capacitor is given an initial charge CE and when the switch S is closed at time $t=0$ current flows from the capacitor discharging it.

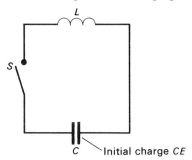

Fig. 2.8

The voltage across the inductor is then $L(di/dt)$ or $L(d^2q/dt^2)$ (since $i=dq/dt$ and $di/dt=d^2q/dt^2$).

The voltage across the capacitor is q/C.
The total applied voltage is zero and it follows that

$$L\frac{d^2q}{dt^2} + \frac{1}{C}q = 0 \tag{18}$$

This equation is similar to (16) and we can state the solution by analogy

L corresponds to m

$\frac{1}{C}$ corresponds to k

$\therefore \quad q = A\cos\omega t$ is the solution.
$A =$ initial disturbance (charge CE in this case)

$$\omega = \sqrt{\frac{1}{LC}} \quad \left(\text{compare } \omega = \sqrt{\frac{k}{m}}\right)$$

or

$$f = \frac{1}{2\pi\sqrt{LC}}\text{ hertz.} \tag{19}$$

2.9 FLUID SYSTEMS

The fluid system produces a similar result to the other two cases.
   The differential equation for some initial displacement of the liquid is

$$I\frac{d^2Q}{dt^2} + \frac{1}{C}Q = 0$$

and its solution is

$Q = Q_0\cos\omega t$ where $Q_0 =$ initial disturbance.

In each of these cases we obtained a differential equation as our mathematical model with a second derivative term d^2/dt^2 in it. Such a differential equation is termed a second-order differential equation and describes a second-order system.
   The systems each had an initial displacement or disturbance and their subsequent behaviour was observed. In each case the solution was of the form $s = A\cos\omega t$. Where on the other hand the system is initially at rest and *then* a disturbance is applied the solution would be of the form $s = A\sin\omega t$. In both instances the result is simple harmonic motion (S.H.M.) but in the first case when $t = 0$, $s = A$ (the initial displacement from the 'rest' position). In the second case when $t = 0$, $s = 0$.

2.10 ENERGY CONSIDERATIONS

The systems described involve the employment of non-dissipative elements, a mass and a spring in the mechanical case, an inductor and capacitor in the electrical case. Consequently no energy is lost (at least not in the idealised case). Energy is merely exchanged between the elements, converting potential to kinetic energy or electromagnetic to electrostatic energy. The elements control that rate at which this energy transfer takes place, that is, the frequency of the exchange.

2.11 THE R/C CIRCUIT USED FOR INTEGRATION OR DIFFERENTIATION

We can link an R and C element together in order to obtain a means of differentiation.

Take the case shown in Fig. 2.9. The input voltage v_i causes a current

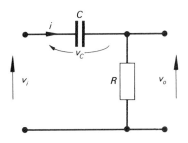

Fig. 2.9

i to flow charging C and producing a p.d. across R.

Let $v_o = $ p.d. across R

and $v_C = $ p.d. across C

Then $v_i = v_C + v_o$

$$= \frac{1}{C} q + Ri = \frac{1}{C} q + R \frac{dq}{dt}.$$

Now if R and C are both small enough

$$\frac{1}{C} q \gg R \frac{dq}{dt} \quad \text{and} \quad v_i \simeq \frac{1}{C} q \quad \text{or} \quad q \simeq Cv_i.$$

Hence $v_0 = R \dfrac{\mathrm{d}q}{\mathrm{d}t}$

$\qquad = RC \dfrac{\mathrm{d}v_i}{\mathrm{d}t}$ (very nearly)

That is, the output voltage is approximately RC times the derivative of the input. This is only so for small time constants.

Interchanging R and C (Fig. 2.10) we get the following

$\qquad v_i = v_C + v_R$

and $v_i = \dfrac{1}{C} q + R i$

$\qquad = \dfrac{1}{C} q + R \dfrac{\mathrm{d}q}{\mathrm{d}t}$ as before

But $v_0 = $ p.d. across C.

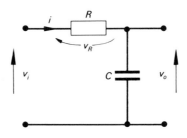

Fig. 2.10

If now R and C are both large enough $R(\mathrm{d}q/\mathrm{d}t) \gg (1/C) q$

$\therefore \qquad v_i \simeq R \dfrac{\mathrm{d}q}{\mathrm{d}t}$ and $\mathrm{d}q = \dfrac{v_i}{R} \mathrm{d}t$

so that

$\qquad q = \dfrac{1}{R} \displaystyle\int v_i \, \mathrm{d}t$

Here $v_0 = \dfrac{q}{C} = \dfrac{1}{CR} \displaystyle\int v_i \, \mathrm{d}t$

That is, the output voltage is approximately $1/RC$ times the integral of the input. This is only so for large time constants.

Input and output waveforms for these circuits are shown in Fig. 2.11.

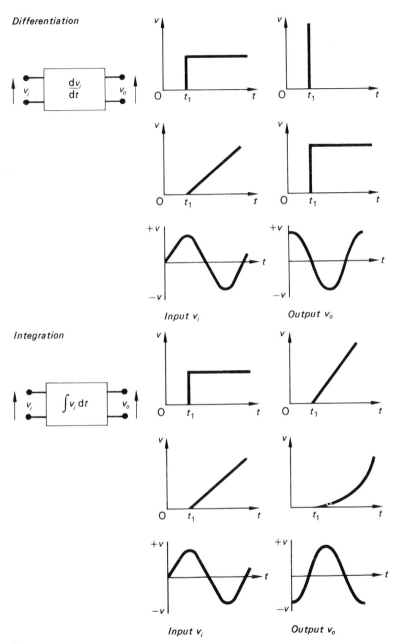

Differentiation

Integration

Input v_i Output v_o

Fig. 2.11

2.12 SUMMARY

The response of first-order systems to a sudden change or stimulus – a step input – is exponential. The response of a second-order system with no damping to a step input is oscillatory.

A first-order system has a time constant which depends upon the parameters of the system.

A second-order undamped system oscillates with simple harmonic motion at a frequency which depends upon the parameters of the system. Certain linked elements have the property of differentiation or integration.

QUESTIONS

2.1 The mass m of 2·5 kg is placed on the horizontal platform. The spring has a compression stiffness of 200 N/m and the dashpot has a damping coefficient of 600 Ns/m.* Find (i) how far the mass has descended in 5 s and (ii) the total distance ultimately descended. What assumptions are made in the calculation?

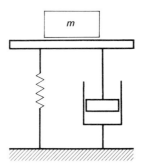

2.2 A first-order system possesses a time constant of 12 s. How long will it take to reach a state corresponding to 50 per cent of the steady state following a step disturbance?

2.3 A mass of 6 kg is suspended from a helical spring of stiffness 500 N/m. If the mass is given a sudden disturbance in a vertical direction find the frequency of oscillation.

2.4 An electrical circuit is required to produce oscillations at 1 MHz. An inductance of 1 mH is available. What capacitance is required across the inductor to achieve this result?

2.5 A rod of negligible mass and length 1·5 m fits over a smooth peg at one end. A mass of 0·5 kg is attached to the lower end. Derive, for small angular displacements θ an expression for the frequency of oscillation. Find the value of this frequency. (Hint: the restoring force $= mg \sin \theta \simeq mg\theta$ for small values of θ.)

2.6 A cylindrical tank of water with a constant head of 5 m and diameter 1·5 m is connected to a similar tank (which is empty) through a small bore pipe which

* Damping coefficient has units of either N per m/s (N/m/s) or Ns/m. These are dimensionally identical.

offers an effective resistance of 0·8 Ns/m⁵. The bases of the two tanks rest on a horizontal plane. Find the time taken for the water in the second tank to reach a height of 1 m.

2.7 A capacitor has a voltage of 150 V applied across it and then stores 0·05 J. It is suddenly connected to an inductor of 0·12 H and negligible resistance. Find the frequency of oscillations produced and the peak value of the charging or discharging current, assuming no loss of energy.

2.8 An integrating circuit consists of a resistor of 2 MΩ and a capacitor of 5 μF. A step unit voltage of 20 V is applied to the two components in series and an output is obtained across the capacitor. Find the difference at the end of 1 s between the ideal output and the actual output.

2.9 The following input voltages (v_i) are applied in turn to the circuit shown.

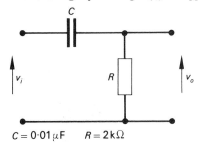

$$C = 0·01\ \mu F \qquad R = 2\ k\Omega$$

Input voltage waveforms

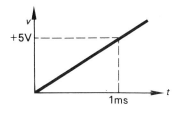

(a) Uniform increase

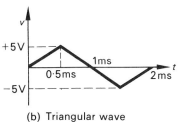

(b) Triangular wave

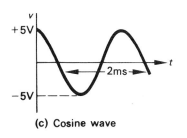

(c) Cosine wave

(d) Parabolic relationship

33

Sketch to scale the output voltage (v_o) assuming the time constant is negligibly small.

2.10 Show from first principles that the temperature rise of an electrical machine operating on constant load is given by

Temp. rise $= \theta_m(1 - e^{-t/T})$ where $\theta_m =$ final temperature rise and $T =$ thermal time constant.

The temperature of a motor operating on constant load is is follows:

Time (h)	0	1	2	4
Temp. (°C)	15	22·9	25·2	26·3

Find the thermal time constant and the final operating temperature. How long after the machine has reached its ultimate temperature will it take for the machine temperature to fall to 20°C after switching off? Ambient temperature remains constant at 15°C.

ANSWERS

2.1 0·0997 m 0·123 m (inertia of mass is ignored) **2.2** 8·318 s
2.3 1·453 Hz   **2.4** 25·3 pF   **2.5** 0·407 Hz   **2.6** 32·15 s
2.7 21·79 Hz 0·913 A   **2.8** 0·097 V

Chapter 3
Second Order Systems

3.1 SECOND-ORDER SYSTEMS WITH DAMPING

So far we have considered second-order systems with no dissipative element. We are now going to add the dissipative component to the equation of motion (or charge).

This is obviously the general situation and the equation for a mechanical second-order system with damping is

$$F = m\frac{d^2s}{dt^2} \qquad + \qquad a\frac{ds}{dt} \qquad + \qquad ks \qquad (20)$$

$$\text{inertia term} \quad \text{damping term} \quad \text{stiffness term}$$

A system which will achieve this is shown in Fig. 3.1.

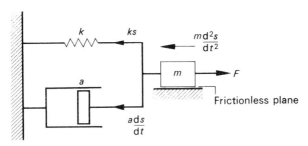

Fig. 3.1

The total force F applied is resisted by the combined efforts of the inertia force, the damping force and the stiffness of the spring.

For an electrical system the three components inductance L, resistance R and capacitance C are connected in series and the voltage produced across them must add up to the applied voltage E.

The differential equation is therefore

$$E = L\frac{d^2q}{dt^2} + R\frac{dq}{dt} + \frac{1}{C}q \qquad (21)$$

While the mathematics needed to solve this equation might be beyond us at the moment we can at least reason out what will happen if a step disturbance is applied. The response will be both exponential and oscillatory. Because of the damping term in the middle of the expression on the right-hand side we can expect the oscillations to die away and a little thought will lead to a solution of the form

$$s = e^{-\alpha t} A \sin \omega_n t$$

or

$$q = e^{-\alpha t} C E \sin \omega_n t \qquad (22)$$

The first part of the solution, $e^{-\alpha t}$, is the factor which controls the rate at which the oscillations die away. The greater the value of α the more rapid will be the decay. ω_n is the undamped natural frequency of the system and depends upon the components employed in the system.

If α is zero there is no damping and we are back to the second-order equation describing S.H.M. If ω_n is zero there is no oscillation.

Figure 3.3 shows the responses obtained in the electrical case when L and C are maintained constant and only R is varied. The graph shows the voltage across the capacitor plotted to a base of time.

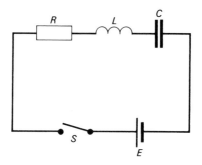

Fig. 3.2

Eventually the step disturbance appears across C but, depending upon R, the response may be oscillatory before this steady voltage is attained. The equation representing the voltage must now have a constant added to it to represent the step disturbance. The frequency of the oscillatory exchange of energy between the inductor and capacitor depends largely upon L and C but is also affected by the resistance (as one might expect since the resistor governs the build up of charge in a first-order system).

The actual value of ω in this general case can be shown to be equal to $\sqrt{(k/m - a^2/4m^2)}$ in the mechanical system and $\sqrt{(1/LC - R^2/4L^2)}$ in the electrical system.

Considering the mechanical system it is apparent that the frequency of the oscillations is less than the undamped natural frequency of the undamped system.

ω obviously is equal to zero when $k/m = a^2/4m^2$, i.e. $a = 2\sqrt{km}$. This is the damping needed to just prevent oscillations and is termed Critical Damping.

The ratio

$$\frac{\text{actual damping}}{\text{critical damping}} = \frac{a}{2\sqrt{km}}$$

is called the Damping Ratio ξ.

ω can now be written as

$$\sqrt{\left\{\frac{k}{m}\left(1 - \frac{a^2}{4km}\right)\right\}} = \sqrt{\left\{\frac{k}{m}(1 - \xi^2)\right\}} \qquad (23)$$

if $\sqrt{(k/m)} = \omega_n$ the undamped angular frequency then

$$\omega = \omega_n\sqrt{(1 - \xi^2)} \qquad (24)$$

The factor α governing the decay rate is given by $a/2m$ (note that a factor of 2 appears in the second-order case).
But

$$a = 2\xi\sqrt{(km)}$$

Hence

$$\frac{a}{2m} = \xi\sqrt{\frac{k}{m}} = \xi\omega_n. \qquad (25)$$

The complete solution for s following a step disturbance F is

$$s = \frac{F}{k}\left[1 - \frac{e^{-\omega_n t}}{\sqrt{(1 - \xi^2)}}\sin\left(\omega_n\sqrt{(1 - \xi^2)}\,t - \varphi\right)\right] \qquad (26)$$

where $\varphi = \cos^{-1}\xi$.

The proof of this solution is outside the scope of this book.

This shows that the damping ratio affects the frequency of oscillations, and the natural frequency in turn affects the decay rate of these oscillations.

3.2 DAMPED AND UNDERDAMPED SYSTEMS

The response of a second-order system (with damping present) to a suddenly applied stimulus (a step function) depends upon the components comprising that system.

A system which has small damping produces a response which is oscillatory, the oscillations finally dying away and the system settles now to a new steady position or state. ξ for such a system is less than unity.

The response just described applied to an 'underdamped system', the response 'overshooting' its final value.

If the damping is greater than critical ($\xi > 1\cdot0$) no oscillations occur and the response of the system follows an exponential law. In practice a wide range of damping ratios are encountered – a ratio of $0\cdot7$ gives an 'overshoot' of 4 per cent during the first cycle of oscillation.

When the damping is critical the system still obeys an exponential law but its time constant is as small as possible without oscillations occurring.

In a simple second-order mechanical system critical damping is given by $a = 2\sqrt{(km)}$.

In an equivalent electrical system the critical damping occurs when $R = 2\sqrt{(L/C)}$. See Fig. 3.3.

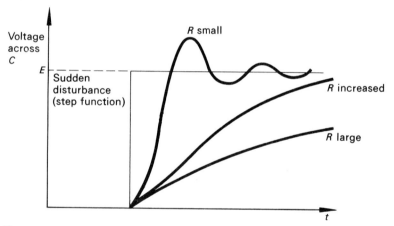

Fig. 3.3

3.3 TIME RESPONSE OF A MECHANICAL SYSTEM

The differential equation for a mechanical second-order system

$$m\,\frac{\mathrm{d}^2s}{\mathrm{d}t^2} + a\,\frac{\mathrm{d}s}{\mathrm{d}t} + ks = F$$

can be rearranged to give

$$\frac{d^2s}{dt^2} + \frac{a}{m}\frac{ds}{dt} + \frac{k}{m}s = \frac{F}{m} \tag{27}$$

or

$$\frac{d^2s}{dt^2} + 2\xi\omega_n\frac{ds}{dt} + \omega_n^2 s = \frac{F}{m} \tag{28}$$

This differential equation must be looked at rather carefully. ξ affects both the decay rate of the oscillations as well as the frequency of the oscillations for underdamped systems. Increasing ξ certainly causes the oscillations to die away more quickly but also *reduces* the frequency of the oscillations. On the other hand if we maintain ξ at a constant value and alter ω_n then increasing this quantity causes the oscillations to die away more quickly and *increases* the frequency of oscillations. For underdamped systems this is a desirable feature since the time response will be reduced and the final value will be reached more quickly.

To increase ω_n we need to increase k or decrease m (or both). We may maintain ξ at a constant value by keeping the product km constant. A system with a large k and a small m is called a 'stiff' system and the response to a step function shows a much greater 'overshoot' of its ultimate value than a system with a lower undamped natural frequency. See Fig. 3.4.

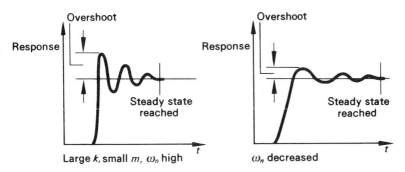

Fig. 3.4

If this 'overshoot' is not acceptable then we have to increase the damping 'a' and with it the value ξ. This in turn will affect the frequency of the oscillations causing the response time to be increased. A compromise is therefore normally forced on us at this stage.

If of course we cannot tolerate any oscillations at all – that is, the system must not 'overshoot' its ultimate value, then we have to aim at critical damping conditions to give the fastest response time.

When critical damping is achieved the solution is

$$a = A[1 - e^{-\omega_n t}(1 + \omega_n t)], \quad A \text{ is the step input.}$$

Even here it will be seen that increasing ω_n causes the system to respond more quickly. There is however a practical limit to the value of ω_n.

It is left to the reader to write down the equivalent equations for electrical and fluid systems and produce the solutions by analogy.

3.4 ROTATIONAL MECHANICAL SYSTEMS

These can probably best be understood by comparison with linear systems.

We have so far discussed linear movement in a mechanical system. The translation to rotational movement is quite straightforward and can be achieved by direct comparison of the relevant parameters of the two modes of movement. In place of a linear displacement s we have an angular displacement θ. See Fig. 3.5.

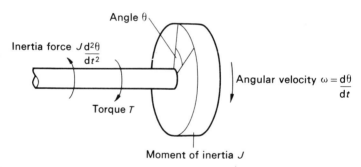

Fig. 3.5

Thus:

Linear motion

$$F = mf = m\frac{\mathrm{d}^2 s}{\mathrm{d}t^2}$$

(m = mass (inertia)

f = linear acceleration in m/s²)

Rotational motion

$$T = J\alpha = J\frac{\mathrm{d}^2\theta}{\mathrm{d}t^2}$$

(T = torque in Nm)

J = polar mass moment of inertia in kgm²

α = angular acceleration in rad/s²

$F = ks$

(k = stiffness in N/m)

$F = a \dfrac{ds}{dt}$

(a = damping coefficient in Ns/m)

$T = K\theta$

(K = torsional or angular stiffness torque in Nm/rad)

$T = C \dfrac{d\theta}{dt}$

(C = angular damping torque Nms/rad)*

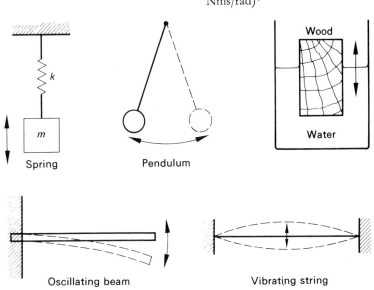

Spring   Pendulum

Wood

Water

Oscillating beam   Vibrating string

The general equation can be written in two forms

$$J \frac{d^2\theta}{dt^2} + C \frac{d\theta}{dt} + K\theta = T_{\text{applied}} \qquad (29)$$

or

$$\frac{d^2\theta}{dt^2} + 2\xi\omega_n \frac{d\theta}{dt} + \omega_n^2\theta = \frac{T_{\text{applied}}}{J} \qquad (30)$$

and there is essentially no difference between this and the linear motion case except that here we are dealing with angular movement.
Thus the undamped natural frequency of oscillation is

$$\frac{1}{2\pi} \sqrt{\frac{K}{J}} \text{ hertz}$$

and critical damping occurs when $C = 2\sqrt{(KJ)}$.

* When dealing with angular damping we can use the units Nm per rad/s (N-m/rad/s) or N-ms/rad. They are dimensionally identical.

3.5 SECOND-ORDER SYSTEMS WHICH OCCUR IN PRACTICE

A number of second-order systems can be observed in everyday life. The balance wheel of a watch is an example. The oscillations follow very nearly a sinusoidal pattern. These oscillations would soon die away completely but for the fact that potential energy stored in the main spring of the watch replenishes the dissipated energy of the system and keeps the balance wheel oscillating. The natural frequency of the balance wheel system (the hairspring and the mass of the wheel) enables accurate time to be maintained.

Another example is the pointer of a direct current ammeter. The system of pointer coil and control spring is damped to prevent 'over-shoot'. Ideally the damping is adjusted to the critical value so that a steady indication is reached as quickly as possible.

A guitar string is an example where a small damping ratio is put to good use. The string has mass and stiffness and is struck or plucked. This is simply applying a step function and the string then responds at its natural frequency giving out a note by the air it displaces. This note is amplified mechanically by the sounding board of the guitar.

A heavy but buoyant mass thrown into a pond probably exhibits second-order motion as it first sinks and then bobs up and down in the water before settling down to its normal floating position. Such factors must be taken into account with the design of ships. A gust of wind on an upright fence may cause oscillations which quickly die away. The effect of wind gusts on buildings such as high-rise flats is a very important factor to be taken into consideration. The stiffness of a building is obviously very considerable but as the inertia is also very large the natural frequency of the building might easily be quite low and could present considerable problems.

3.6 SUMMARY

First-order systems are characterised by first-order differential equations resulting in an exponential response to a step disturbance.

Second-order systems are characterised by second-order differential equations and the response depends upon the degree of damping. With zero damping the response is sinusoidal and oscillations continue at the undamped natural frequency with uniform amplitude. With increasing damping the oscillations die away exponentially and their frequency is below that of the undamped natural frequency. With critical damping the oscillations are just prevented from occurring.

Rotational mechanical second-order systems behave in a very similar fashion.

QUESTIONS

3.1 A mechanical second-order system consists of a spring (stiffness 20 N/m) and a mass 0·5 kg. What is the natural frequency? What value of damping is needed to give critical damping? Sketch the response of the system (i) without damping (ii) with critical damping.

3.2 An electrical circuit consists of a capacitor 0·2 μF, an inductor 0·1 H and a variable resistor in series with a switch. The capacitor is initially charged. Sketch the waveform of current after the switch is closed if the value of resistance is less than the value needed for critical damping. If the damping ratio is 0·1 what is the frequency of oscillation? What value of resistance is needed to achieve this?

3.3 Find the required mass m in the system shown to produce critical damping ($k=50$ N/m, $a=25$ Ns/m). What is the final value of the spring extension and how far will it extend in 0·25 s from the instant that the mass is placed on the platform, if the spring is initially unextended?

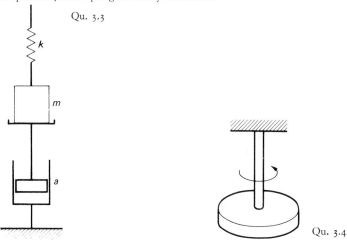

Qu. 3.3

Qu. 3.4

3.4 A solid cylindrical mass is attached to the end of a shaft of angular stiffness torque 8 Nm/rad. The mass has the dimensions diameter, 200 mm, width 20 mm and has a density of 7 000 kg/m³. Find the frequency of torsional oscillations. What assumptions (if any) have been made? Describe a practical method of damping out these oscillations and find the minimum value of angular damping torque to prevent oscillation.

3.5 A railway truck mass m moving with velocity u along a level track strikes some spring loaded buffers with compression stiffness s. Write down the force equation and sketch a graph of velocity of the truck against t. If the motion of the buffers is now damped with damping constant a produce a new force equation. If the damping is adjusted to the critical value find the maximum compression of the buffers.

ANSWERS

3.1 (i) 1·006 Hz, (ii) 6·324 Ns/m   **3.2** 141·4 Ω   1120 Hz   **3.3** $m=3·125$ kg
$s=0·613$ m   0·162 m   **3.4** 3·036 Hz   ~~3·52~~ Nm/rad/s
0·84

43

Chapter 4
System Response

4.1 BLOCK DIAGRAMS AND EQUIVALENT SYSTEMS

Rather than actually drawing springs and dashpots or inductors and resistors we can show the elements of a system using a rectangular block and indicating what is in the block by either labelling it or giving the mathematical expression which describes it. The mathematical expression is called the transfer function.

For example we can consider a mass, a spring and a dashpot all being connected to a bar. The block diagram illustrating this is shown in Fig. 4.1. The electrical system of inductance, capacitance and resistance in series can also be shown by a block diagram and the labels

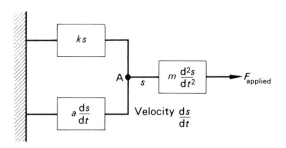

Fig. 4.1

L, C and R should suffice. In comparing these two systems (which are known to be similar) one point should now be obvious – the forces are acting side by side, i.e. in parallel and the voltages are acting in line, i.e. in series. Thus forces in parallel are like voltages in series in the systems studied. Conversely forces in series are like voltages in parallel in similar situations. We can use this concept in the study of more complex systems.

One other point of significance. Point A on the mechanical system is common to the mass, spring and dashpot. Hence the velocity of the point A, ds/dt, is the same for all elements.

In the electrical system, Fig. 4.2, the charge q is the variable quantity rather than distance s.

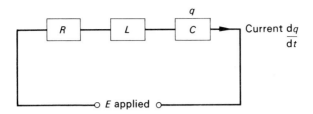

Fig. 4.2

Now $dq/dt = i$ (current) and dq/dt is analogous to ds/dt, the velocity. It follows therefore that in this system current *through* the three elements L, C, R is the common factor.

Current *through* the *series* electrical elements is similar to the *velocity* of a point in the *parallel* mechanical elements for the systems described.

Consider the spring mass system shown by a block diagram (Fig. 4.3) which represents two springs in series. Assume that the force F applied to the mass causes the point A to move downwards a distance s_1. This in turn causes the upper spring to move a distance s_2 (point B).

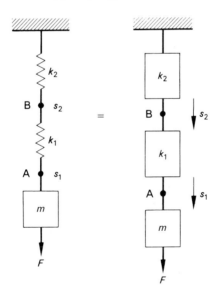

Fig. 4.3

Thus the restoring force produced by the lower spring is $k_1(s_1 - s_2)$.

The equation of force on the mass m:

    Applied force = Inertia force + stiffness (restoring) force

i.e.

$$F = m\frac{d^2 s_1}{dt^2} + k_1(s_1 - s_2) \qquad (31)$$

But the force produced by the lower spring must be equal to the force produced on the upper spring.

Thus $k_1(s_1 - s_2) = k_2 s_2$

or $k_1 s_1 = s_2(k_2 + k_1)$

hence $s_2 = \dfrac{k_1 s_1}{k_2 + k_1}$

Substituting in equation (31)

$$F = m\frac{d^2 s_1}{dt^2} + \frac{k_2 k_1}{k_2 + k_1} s_1 \qquad (32)$$

The two springs in series have a new stiffness $k_1 k_2/(k_1 + k_2)$. We can draw out equivalent series electrical circuit – Fig. 4.4.

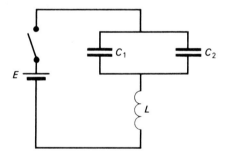

Fig. 4.4

Now if $k_1 = 1/C_1$ and $k_2 = 1/C_2$

$$\frac{k_1 k_2}{k_1 + k_2} = \frac{1/C_1 C_2}{1/C_1 + 1/C_2} = \frac{1/C_1 C_2}{(C_1 + C_2)/C_1 C_2} = \frac{1}{C_1 + C_2} \qquad (33)$$

The equivalent capacitance is therefore $C_1 + C_2$ obtained by two capacitors in parallel.

The two capacitors C_1 and C_2 in *parallel* are equivalent to the two springs k_1 and k_2 in *series*.

We can easily find the undamped natural frequency of these new systems by inspection:

$$f_n = \frac{1}{2\pi\sqrt{\{L(C_1+C_2)\}}} \quad \text{Hz} \quad \text{for the electrical system}$$

$$f_n = \frac{1}{2\pi\sqrt{\left\{m\left(\dfrac{k_1+k_2}{k_1k_2}\right)\right\}}} \quad \text{Hz} \quad \text{for the mechanical system}$$

Consider the electrical circuit Fig. 4.5. What is the equivalent mechanical system? Here the two capacitors are in series, equivalent to two springs in parallel. The mechanical system is that shown in Fig. 4.6.

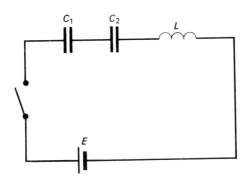

Fig. 4.5

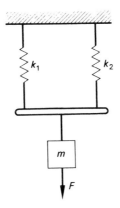

Fig. 4.6

The undamped natural frequency of the electrical circuit is

$$f_n = \frac{1}{2\pi \sqrt{\left\{ L \left(\frac{C_1 C_2}{C_1 + C_2} \right) \right\}}}$$

since capacitors in series have a total capacitance given by

$$\frac{1}{C_{total}} = \frac{1}{C_1} + \frac{1}{C_2}$$

i.e. $\quad \dfrac{1}{C_{total}} = \dfrac{C_2 + C_1}{C_1 C_2} \quad$ and hence $\quad C_{total} = \dfrac{C_1 C_2}{C_1 + C_2}$

The undamped natural frequency of the mechanical circuit

$$f_n = \frac{1}{2\pi \sqrt{\{ m/(k_1 + k_2) \}}} \ \text{Hz} \tag{34}$$

Let us consider a practical application. The suspension of a wheel of a train, Fig. 4.7, shows the basic requirements consisting of two springs

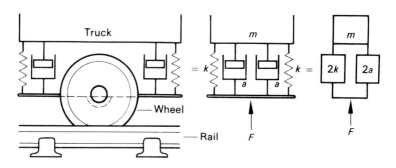

Fig. 4.7

connecting the axle to the truck. As this system would be liable to oscillation if any slight disturbance occurred on the track, dampers are necessary. These would be oil filled to create a large damping coefficient. The two springs in parallel have an effective stiffness of $2k$ and the two dashpots produce a damping force $2a \times$ vertical velocity. The system can therefore be simplified. A little thought will show that our final model may be a little oversimplified as it is possible to produce a rocking action on this type of suspension where one spring may be extended more than the other.

4.2   MORE COMPLEX SYSTEMS

Let us now consider a system with three springs and two masses as illustrated in Fig. 4.8.
If we give m_1 an initial displacement what happens?

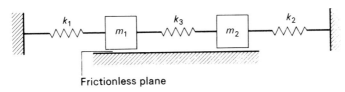

Fig. 4.8

Let m_1 move a distance s_1 to the right and m_2 move a distance s_2 to the left after the application of F.
For the mass m_1

$$m_1 \frac{d^2s_1}{dt^2} = \text{inertia force}$$

Now the spring forces are due to k_1 and k_3.
k_1 is extended s_1 but k_3 is compressed $s_1 + s_2$

$$\therefore \qquad F = m_1 \frac{d^2s_1}{dt^2} + k_1 s_1 + k_3(s_1 + s_2) \qquad (35)$$

Similarly for mass m_2

$$0 = m_2 \frac{d^2s_2}{dt^2} + k_2 s_2 + k_3(s_1 + s_2) \qquad (36)$$

Obviously this is a much more complicated situation with one mass spring system affecting the other through the common spring k_3. Each has a natural frequency of oscillation but when connected two natural oscillation frequencies of the whole system are possible neither equal to the natural frequency of one spring mass system acting on its own.

Even when $m_1 = m_2$ and $k_1 = k_2$ there are still two natural frequencies of the system, the lower when m_1 and m_2 move in the same direction and the centre spring is neither extended nor compressed. The higher natural frequency occurs when m_1 and m_2 are moving symmetrically in opposite directions.

The electrical analogy is shown in Fig. 4.9 which some students might find easier to understand.

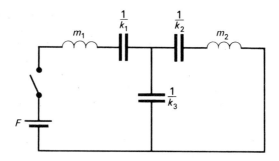

Fig. 4.9

If the coupling spring k_3 has a very small stiffness and $m_1 = m_2$ and $k_1 = k_2$ then if one system is set in oscillation the energy will be gradually transferred to the other and the first spring mass system will come to rest. The energy transfer will then occur in the opposite direction. A system so described is a second-order system with two degrees of freedom as opposed to the other systems dealt with so far which were systems with one degree of freedom. It is suggested that systems with two degrees of freedom might make a useful series of practical investigations in a qualitative manner.

It is not proposed in a book at this level to investigate such systems in an analytical fashion. Rather more mathematics is required since the equations of force involve the solution of second-order simultaneous equations.

4.3 STIMULUS AND RESPONSE

A group of linked elements in a system respond in varying ways to different stimuli. The type of stimulus that we have dealt with so far has been limited to a sudden change – a step function. In the electrical case this was achieved by closing a switch and suddenly applying a voltage or discharging a capacitor. In the mechanical case a force which had displaced a system was suddenly removed. The response to step functions of the systems we have dealt with was always one of exponential growth (or decay) or oscillation or a combination of the two.

The step function stimulus may be called the 'forcing function'. The response of the system was such that it eventually settled down to a new 'steady state'. In the case of the spring mass, damping system when a force initially displacing the system was removed, the mass eventually settled down in a new position. In the electrical system of L, C and R in series, a suddenly applied voltage E caused the system to oscillate with decreasing amplitude until the new 'steady state' was one where the charge on the capacitor (initially zero) rose to CE.

In each case the system changed from one stable situation to another but in the transition between the two stable states the system responded in an oscillatory or exponential way. This transition period is known as a 'transient'. Transients in a stable system eventually die away. The response of a system to a step function is essentially a transient.

In general

$$\text{Total response} = \text{Transient response} + \text{steady state response}$$

Initially before the steady state is reached the transient controls the behaviour of the system. After some time the transient disappears and the steady-state condition is reached.

4.4 SINE FORCING FUNCTIONS

Systems may be subjected to stimuli other than step functions. One of the commonest alternatives is the sine or cosine function. The electrical engineer has some advantage over his mechanical colleague here since he is (or at least should be!) very familiar with the effect of the application of sinewave voltages to L, C, R, circuits. This comes under the general heading of a.c. theory.

For example, we may wish to know the current i through a series circuit consisting of an inductance L henrys in series with a resistance R ohms when a voltage $v = V \sin \omega t$ is applied. We use rotating vectors (phasors) to describe v and i and we end up with a phase angle φ between them. Thus $i = V/Z \sin(\omega t - \varphi)$ where Z, the impedance $= \sqrt{\{R^2 + (\omega L)^2\}}$ and $\varphi = \tan^{-1} \omega L/R$ by normal a.c. theory methods. See Fig. 4.10. This result is essentially the 'steady state' situation, any transient effect having been assumed to have died away. In a.c. circuits where ω is usually of the order of several hundred radians per

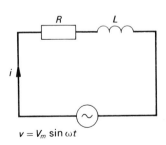

$v = V_m \sin \omega t$

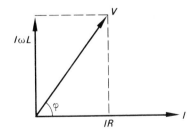

Fig. 4.10

second or more, we can only observe the transient on a cathode-ray oscilloscope since it usually lasts for only a very brief period.

In mechanical systems where the transient may have a longer time constant and where the natural frequencies are often much lower we must be always aware that the transient may take an appreciable time to decay to negligible proportions.

In all the electrical systems dealt with we have used q rather than i as the variable quantity. We know that $i = dq/dt$ and as it follows that $q = \int i \, dt$, we can therefore obtain q easily from the relationship for i. In the example above $i = V/Z \sin(\omega t - \varphi)$

Hence

$$q = \frac{V}{Z} \int \sin(\omega t - \varphi) \, dt = \frac{-V}{\omega Z} \cos(\omega t - \varphi)$$

$$\text{or} \quad \frac{V}{\omega Z} \cos(\omega t - \varphi + \pi) \tag{37}$$

4.5 RESONANCE

In general with a series L, C, R circuit for a voltage $v = V \sin \omega t$ applied the current is given by

$$i = \frac{V}{Z} \sin(\omega t - \varphi)$$

where

and

$$\left.\begin{array}{c} Z = \sqrt{\left\{ R^2 + \left(\omega L - \dfrac{1}{\omega C} \right)^2 \right\}} \\[2em] \tan \varphi = \dfrac{\omega L - 1/\omega C}{R} \end{array}\right\} \tag{38}$$

Remember that we are concerned here with only 'steady state' conditions, the transients which exist have died away before we make our observations.

A special case occurs when $\omega L = 1/\omega C$ – this is a *resonant* condition. When this happens $Z = R$ (a minimum value) and $\phi = 0$. If therefore we plot i against frequency f for such a circuit with a constant amplitude V of voltage applied we find that the current through the circuit increases as we approach resonance and passes through a maximum value. Since resonance occurs when $\omega L = 1/\omega C$ it follows that the resonant angular frequency ω_0 is given by

$$\omega_0^2 = \frac{1}{LC} \quad \text{or} \quad \omega_0 = \frac{1}{\sqrt{LC}} \quad \text{and hence} \quad f_0 = \frac{1}{2\pi} \frac{1}{\sqrt{LC}} \text{ Hz}$$

It should be noted that this is the same as the natural frequency of a circuit possessing negligible resistance but is slightly higher than the natural frequency when resistance is present.

If the resistance of the circuit is small then the current at resonance can be large – much larger than at 'off resonance'. The graph i/f is shown in Fig. 4.11. The corresponding phase angle between v and i also varies with frequency. This is also shown in Fig. 4.11.

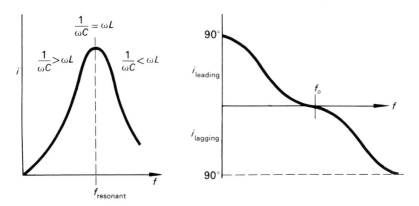

Fig. 4.11

4.6   MECHANICAL SYSTEMS SUBJECTED TO SINEWAVE FORCING FUNCTIONS

Mechanical circuits respond in much the same way as electrical circuits and if we apply a sinusoidal force to a mass, spring, damper system the response is very similar. We can even use a concept of mechanical impedance, see Fig. 4.12. (Here the rotation of the wheel causes the force to vary in a sinusoidal manner, i.e. $F = F_m \sin \omega t$.)

The mechanical impedance $Z_m = \sqrt{\{a^2 + (\omega m - k/\omega)^2\}}$ by analogy. Conditions at which Z_m is a minimum occur when $\omega m = k/\omega$ or

$$\omega_0 = \sqrt{\frac{k}{m}} \text{ rad/s} \quad \text{and hence} \quad f_0 = \frac{1}{2\pi} \sqrt{\frac{k}{m}} \text{ Hz}$$

The 'mechanical current' is $F_m/Z_m \sin(\omega t - \varphi)$

where $\varphi = \tan^{-1} \left(\dfrac{\omega m - k/\omega}{a} \right)$

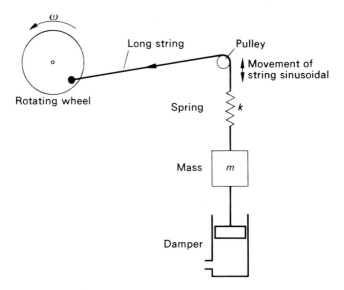

Fig. 4.12

Mechanical current is analogous to dq/dt and from the previous work this corresponds to the velocity of the system. Hence velocity is given by $F_m/Z_m \sin(\omega t - \varphi)$.

It is obvious that at resonance the velocity reaches a maximum and if the damping 'a' of the system is small, dangerously high velocities will be achieved. Resonance is normally to be avoided in mechanical systems. We must guard against it and ensure that if any mechanical structure is going to be subjected to an oscillating force, then its resonant frequency must be well removed from the frequency of the applied force. If there is any possibility of resonance in some way or other then the system must be well damped to avoid dangerously high velocities (and displacements) being produced.

Massive structures have been known to be destroyed by comparatively small applied forces which set up resonant conditions. The Tacoma Narrows Bridge disaster in 1940 is a well known example of unwanted resonance.

4.7 FILTER SYSTEMS

It is possible to consider quite complex systems from a steady-state point of view, especially when the forcing function is sinusoidal.

Take for example the electrical circuit shown in Fig. 4.13. This is a

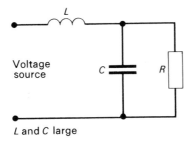

L and C large

Fig. 4.13

circuit which allows low frequency components from the voltage source to be applied to R but provides a high impedance to high frequencies.

It is in fact a filter circuit and because it passes low frequency components more easily than high frequency it is termed a low pass filter.

It is possible by normal phasor methods to find the ratio of the voltage across R and the supply voltage at any frequency. We can produce a similar mechanical system as shown in Fig. 4.14.

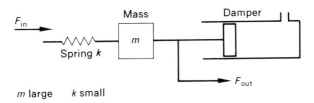

m large   k small

Fig. 4.14

This system will behave in a similar manner to the electrical case and high frequency oscillations will be damped out. Such an arrangement might feature in a form of anti-vibration mounting.

Again take for example a fluid system. A compressor maintains the air pressure of a system at a steady value. The pump action unfortunately generates an unwanted sinusoidal disturbance in the system. How can this be removed?

A filter system is required, see Fig. 4.15. This consists of a narrow pipe connecting a large pressure vessel to the compressor. Another similar vessel is connected to the first. The electrical equivalent is shown in Fig. 4.16. Again this forms a low pass filter severely attenuating the unwanted sinusoidal disturbance while not disturbing the steady pressure.

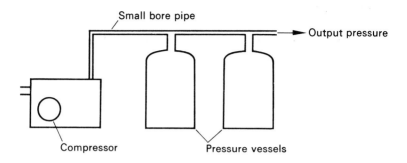

Fig. 4.15

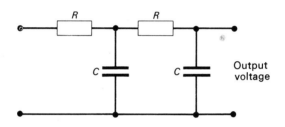

Fig. 4.16

4.8 SUMMARY

First- and second-order systems may be subjected to a variety of stimuli, the most common which are normally considered are the step disturbance and the sinusoidal disturbance. In each case the system response consists of a transient and steady-state component. Normally the transient quickly dies away and the system responds to the forcing disturbance. In the case of sinusoidal forcing functions the system can be dealt with using concepts of impedance, phase angles and phasors. Series resonance can cause very violent responses and is often an undesirable condition especially in mechanical systems.

It is possible to set up analogous mechanical and electrical systems with, in general, mechanical components in parallel corresponding to electrical circuits in series. Filters may be produced in both electrical and mechanical systems.

QUESTIONS

4.1 Draw the equivalent electric circuit. Is the system over, under or critically damped?

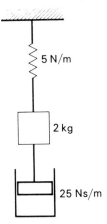

4.2 $a = 50$ Ns/m, $m = 1$ kg, $k = 20$ N/m. Draw the equivalent electrical circuit indicating values.

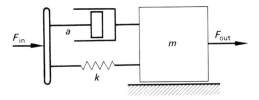

4.3 Show that these two systems are equivalent. If $k_1 = 5$ N/m and $k_2 = 15$ N/m and $m = 50$ kg, find the values of L, C_1 and C_2, and hence find the natural frequency of the system.

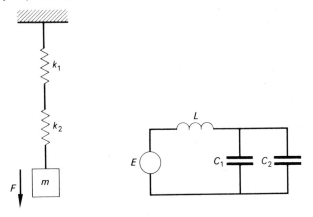

4.4 An electrical circuit consists of an inductance of 0·5 H and a resistance 120 Ω and a capacitance of 25 μF in series. A voltage $v = 100 \sin \omega t$ is applied. Find the equation for the current. What additional value of capacitance is needed for resonant conditions and what is then the current flowing?

4.5 A filter circuit consists of an inductor (20 H, 100 Ω) in series with a capacitor of 50 μF. A voltage $120 + 15 \sin 500\, t$ is applied to this filter and the output is taken across the capacitor. Find the percentage of the a.c. component of voltage to the d.c. component of voltage across the capacitor.

ANSWERS

4.1 Overdamped, L and R practical, C not. **4.3** 0·0436 Hz.

4.4 $i = 0·309 \sin (500\, t + 68·2°)$, 55 μF, 0·833 A **4.5** 60·2 mV 0·05 per cent

Chapter 5

Transducers and Transformers

5.1 INTRODUCTION

Originally transducers were considered to be devices which translated energy from one form to another. They were moreover reversible in their action and could transform the energy either way. Nowadays the term is used rather more loosely (especially by systems engineers) and the following definition will be employed in the text.

5.2 DEFINITION

Transducers are devices for producing one variable quantity in terms of another where the two variables are in different energy systems.

For example the aneroid barometer is essentially a transducer where changes in air pressure on a sensitive bellows (a fluid system) results in the rotation of a pointer (a mechanical system). A pressure variable is transformed into a torsional variable. If the translation is mathematically linear then the change in angular position is directly proportional to the change in pressure, i.e. $\theta \propto$ pressure.

Many transducers measure non-electrical quantities which are then translated to electrical quantities. The reason for doing this is because electrical quantities can be measured accurately usually without recourse to elaborate equipment and electrical signals can be transmitted over distances usually more easily than non-electrical.

5.3 AN EXAMPLE

A strain gauge is an example of such a transducer. Strain is the ratio of the extension of a sample of material over its original length when that sample is subjected to stress. Often the extension is very small and if a direct measurement is needed then some optical or mechanical amplification of the extension is needed. This involves the use of an expensive piece of equipment (the extensometer) which must be placed on the material undergoing tests and the person making the measurement must be situated close to the apparatus.

A strain gauge is essentially a length of thin nichrome or cupronickel wire arranged in a 'W' formation which is cemented to the sample undergoing stress. See Fig. 5.1. The strain gauge's resistance changes

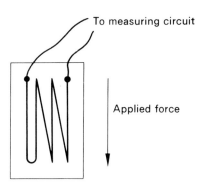

To measuring circuit

Applied force

Fig. 5.1

as the wire is subjected to stress at the same time as the material being tested. The change in resistance (usually measured on a Wheatstone bridge) is an indication of the strain. The bridge may be situated some distance from the gauge and in certain circumstances the strain gauge may be the only means of making the measurement (e.g. measurements on the wing of an aircraft in flight). It is essentially an indirect method of measurement. Sensitive gauges produce a fairly large change in resistance for a given strain but suffer from a major defect in their temperature coefficient of resistance which may also produce changes in resistance. This defect is usually overcome by the use of a second identical strain gauge placed near the first but not subject to any stress. On the assumption that the temperature of the unstressed gauge is the same as the stressed gauge, resistance changes due to temperature changes cancel out (see Fig. 5.2).

It is necessary to calibrate the strain gauge or at least know from the manufacturer of the gauge the change in resistance for a given strain. This information is called the Gauge Factor.

$$\text{The gauge factor} = \frac{\text{Percentage change in resistance}}{\text{Percentage strain}}$$

$$= \frac{100 \times \delta R / R}{100 \times \delta l / l}$$

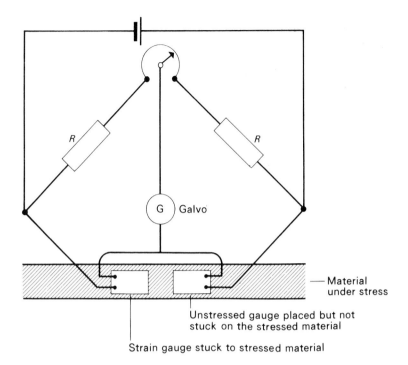

Fig. 5.2

R refers to the original resistance and δR to its change

l refers to the original length of the sample under test and δl to its change.

A typical gauge factor is 2·2. Hence percentage strain = percentage change in $R \div 2\cdot2$.

If the strain is non-static, i.e. is dynamic or changing, then the method of measurement often employed is to balance the Wheatstone bridge initially and then to record the 'out of balance' voltage across the galvanometer terminals on an oscilloscope. The strain gauge is essentially an electrical method of measuring displacement.

5.4 MEASUREMENT OF THE RATE OF CHANGE OF DISPLACEMENT ds/dt

The quantity ds/dt is a velocity and the method of measurement employed depends to a large extent upon the magnitude of the velocity.

We can measure the time taken to travel a given distance and this will then give the average velocity over the given distance. We might wish to know the speed of a rifle bullet for example or the speed of a car. Obviously these are two quite different problems, the method employed in the former where speeds around 600 m/s are expected would not be suitable for vehicle speeds nearer 30 m/s. Moreover the problem of the rifle bullet is very much a 'one-off' situation, whereas often in the car we wish to know the speed at any instant – a continuous monitoring of the speed.

With the first measurement of speed, the bullet may be caused to pass between a pair of light sources and photoelectric cells a fixed distance apart as shown in Fig. 5.3. This will generate two pulses of

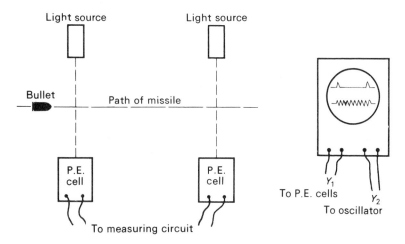

Fig. 5.3

electric current or voltage which after suitable amplification can be applied to one beam of a double-beam oscilloscope. The first pulse can then trigger off a time base which causes the spot on the screen to move across the screen at high speed. The other beam can be connected to a crystal controlled oscillator operating at precisely 10 kHz. The response is photographed. Now by counting the number of complete cycles on one beam between the 'start' pulse and the 'stop' pulse it is easy to calculate the required time interval as each cycle of the oscillations corresponds to 0·1 ms. Hence if the photoelectric cells are 2·5 m apart and the number of cycles counted on the oscilloscope is 45 then the average velocity of the rifle bullet over this distance is $2·5/45 \times 0·1 \times 10^{-3} = 550$ m/s.

It is nowadays fairly easy to count the number of cycles which have occurred over a short interval by electronic means. If the oscillations occur at 1 MHz it is possible to measure the time interval to ± 1 μs allowing an accuracy of two orders beyond that indicated by the oscilloscope method.

The second problem involves the measurement of a much smaller speed but this time a continuous measurement is usually required. The speed of the car is directly related to the rotational speed of the wheels. If the revolutions per minute are measured then knowing the road wheel diameter we have the speed of the vehicle. The rotational speed is measured by coupling a small d.c. generator to the wheels (often via a flexible drive and a gear box). The emf generated by a d.c. generator is directly proportional to the rotational speed if the field flux is maintained constant. Therefore by using a small permanent magnet to generate the magnetic field the emf from the rotating armature may be measured by a high resistance voltmeter and the indication is proportional to the speed of the vehicle. Such a device is referred to as a tachogenerator.

5.5 MEASUREMENT OF ACCELERATION

Since acceleration is the rate of change of velocity any system which produces an emf proportional to speed can be easily adapted to measure acceleration. The velocity emf is simply applied to a differentiating circuit (capacitor in series with a resistance) and the voltage across the resistance is approximately proportional to the acceleration (the smaller the time constant the more accurate is the result).

An alternative method is to use the fact that since force $F = \text{mass} \times$ acceleration then acceleration is directly proportional to the force produced on a given mass. The force may be measured by the extension of a spring and the extension of the spring in turn can be measured by a number of means, if need be by purely mechanical methods. Strictly speaking if the extension is measured by mechanical means a transducer is not involved – there is no change of energy system. Devices which measure acceleration are termed accelerometers.

We can draw transducers as rectangular boxes in our diagrams showing the input variable between two 'terminals' and the output variable (usually electrical) between two other terminals as shown in Fig. 5.4.

5.6 OTHER FORMS OF TRANSDUCERS

A whole range of ingenious transducer devices exist for measuring mechanical, fluid, heat and chemical parameters and transforming

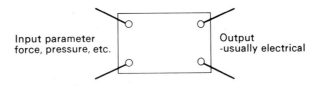

Input parameter
force, pressure, etc.

Output
-usually electrical

Fig. 5.4

them into electrical parameters. The thermocouple is an excellent
example of a heat transducer, where an emf is produced when a
temperature difference is maintained between a pair of junctions of
dissimilar metals in series (see Fig. 5.5). It is as well to handle a number
of transducers, to find out how they function and to measure their
characteristics and relative sensitivities.

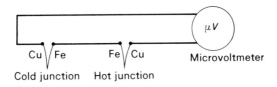

Cu / Fe   Fe / Cu   Microvoltmeter

Cold junction   Hot junction

Fig. 5.5

Two devices worthy of mention are the microphone and the loud-
speaker. In the former we are changing from a fluid system (air pressure)
to an electrical system. In the second the transducer transforms electrical

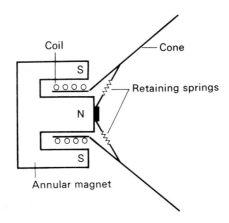

Coil   Cone

S

N   Retaining springs

S

Annular magnet

Fig. 5.6

signals into changes of air pressure. The variation of current in the coil of the speaker causes the lateral force on the coil to vary and the coil moves against the action of a spring. The coil is attached to a stiff cone which in turn causes changes in air pressure and generates sound (see Fig. 5.6).

The reverse action – air pressure causing some electrical disturbance – as occurs in the microphone can be achieved in a variety of ways. In one case a flat rectangular crystal of quartz is clamped at one end. When the other end is subjected to a varying force a small emf is generated across the two faces of the crystal, the polarity and magnitude depending upon the direction and strength of the force applied. A variation of this system where a diaphragm is attached to the free end of the crystal enables pressure to be measured.

5.7 SENSORS, TRANSMITTERS AND ACTUATORS

SENSORS

Sensors are essentially transducers where the position of an object is related to some output quantity – usually an electrical output.

The linear or angular position of the object can be measured by the use of a simple potentiometer.

Take for example the linear potentiometer shown in Fig. 5.7. If a steady voltage is maintained across the ends a, b, of the resistor the

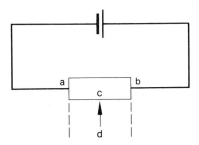

Fig. 5.7

voltage appearing between a and c is directly proportional to the distance of the arm d from one end. If this arm is attached to the object whose position is required then the output voltage is proportional to distance of the object from some reference. By bending the resistor into the shape of a circle the angular position of the arm is directly proportional to the voltage appearing between one end of the resistor and the wiper arm. For example if we maintain precisely 36 V across the ends of the resistor then every volt corresponds to an angle of 10°. Potentiometers suffer from one major disadvantage, namely the wear caused by the continual wiping action of the contact, which may in time cause an electrical breakdown.

One way of avoiding this is to use a differential transformer (Fig. 5.8).

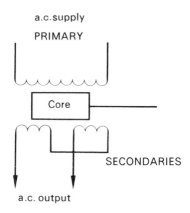

Fig. 5.8

An alternating voltage is maintained across the large winding and the outputs of the two secondaries are in opposition. When the core is in the mid position the coupling from the main primary windings to the secondaries is the same. Two equal voltages appear across the two secondaries but due to the mode of connection the total output voltage is zero.

Displacement of the core from its central position disturbs the voltage balance and the secondary voltage is no longer zero. If two rectifiers are employed as shown in Fig. 5.9 the d.c. output voltage varies in magnitude and polarity with the position of the core. If the core is placed on a spindle then the sensor detects angular position.

Sometimes the reverse procedure is needed, that is the position of a device is an indication of some quantity. The pointer of a moving coil ammeter is a simple and obvious example where the current through the coil produces a torque acting against the action of a control spring. The system attains a steady angular position dependent upon the current flowing. The angular position is shown by a pointer attached to the coil.

One form of fluid sensor has already been mentioned at the beginning of this chapter – the aneroid barometer. A modification of this is a

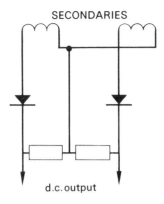

SECONDARIES

d.c. output

Fig. 5.9

differential bellows (Fig. 5.10). The position of the pointer arm is dependent upon the difference between the two pressures *A* and *B*.

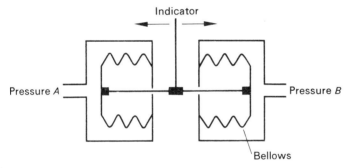

Indicator

Pressure A

Pressure B

Bellows

Fig. 5.10

Another form of sensor which is used extensively in pneumatic control systems is the flapper and nozzle (Fig. 5.11). Air at a steady pressure is passed through the restriction at a, passes down the tube, some continuing on to the nozzle the remainder to the output. Providing the nozzle is unobstructed the air pressure between it and the restriction is low and the output pressure P_0 is small. If the flapper is placed near the nozzle obstructing the air flow the pressure in the tube increases and the output pressure increases.

In practice the flapper is connected to some position sensing element so that the output pressure is a function of the position of the flapper.

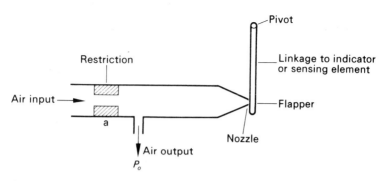

Fig. 5.11

TRANSMITTERS

The linear or angular position of an object may be information that has to be transmitted over a considerable distance. The object might not be easily accessible and some form of remote indication may be necessary. Obviously we need to make use of the information given to us by the sensor. Applying the output of a potentiometer to a high resistance sensitive voltmeter is one means of remote indication of position – the meter would be rescaled in degrees or metres. There is obviously a practical limit to the distance over which such information may be sent.

The system also suffers from two major disadvantages

1. The load imposed on the potentiometer by the voltmeter and its leads. This results in a distortion of the scale of the instrument.
2. The voltage across the potentiometer has to be maintained constant at some reference level.

A device which has been employed frequently in remote indication is the 'synchro'. It is used to indicate angular position. It consists essentially of two identical units each of which closely resembles a synchronous motor. See Fig. 5.12. The stator of the transmitter (or receiver) is a three-phase distributed winding. The rotor is a single phase winding brought out to flexible leads (or if more than 360° rotation is needed to slip rings). Corresponding ends of the stator winding are connected and the two rotors are connected to the same supply (50 Hz mains if available).

The magnetic field from the transmitter rotor circuit induces emf's at different magnitudes and phase in the three parts of the stator winding

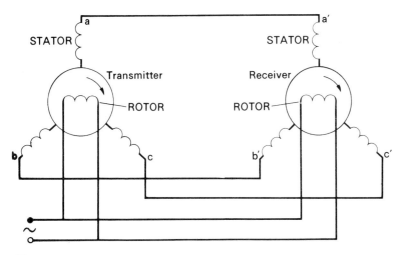

Fig. 5.12

and because these are connected to the receiver synchro the emf's cause circulating currents in the receiver stator.

The receiver therefore sets up a magnetic field which in turn generates a torque on the energised rotor of the receiver. This causes the rotor to turn until it aligns with the magnetic field.

If the rotor of the transmitter is therefore turned through an angle – displacing the magnetic field in the receiver the rotor of the receiver will turn through the same angle. It is as if the two rotors were mechanically coupled.

This method of remote indication can only be used to transmit small torques and if the receiver rotor is loaded in any way insufficient torque may be generated to turn the rotor.

It is possible to overcome this defect by the use of a servo motor. Here any misalignment between transmitter and receiver rotor causes a voltage to appear at the receiver. This can be suitably amplified and applied to a motor which drives the rotor shaft towards the correct angle. When the receiver and transmitter are in line no voltage appears at the receiver. This is essentially an example of a control system which will be dealt with in Chapter 7.

ACTUATORS

It is often necessary in a system to move a large mass through an angle or a linear distance. In order to be able to do this some form of actuator is required. The two main requirements of an actuator are (i) it must produce sufficient power and (ii) it must be controllable.

One form of actuator is a simple d.c. motor which is attached to a suitable gear box such that the angular velocity of the output is smaller than that of the input.

If an angular movement of a large mass is needed (e.g. a crane) then the direction and speed of the motor has to be controlled. This is simply achieved by using a d.c. motor with a separate field winding. The field is energised by a steady current. Passing a varying direct current through the armature enables the speed of the armature to be governed. Reversing the direction of the armature reverses the direction of rotation.

Another form of actuator is the magnetic clutch. The power from a driving motor running at constant speed is coupled via a suitable gear box to the required shaft through a clutch mechanism. The clutch is engaged only when the coil is energised. When the driven shaft is in the required position the clutch is disengaged (Fig. 5.13).

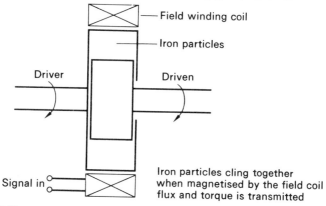

Fig. 5.13

5.8 TRANSFORMERS

Unlike transducers, transformers do not change energy from one form to another. They merely change the magnitude of a parameter within the same energy system and they are reversible in their operation. They are furthermore 'passive' elements in that the output parameter possesses no more energy than the input parameter. Indeed due to the fact that the transformer cannot be 100 per cent efficient the useful output power is always less than the input power. They serve a number of useful purposes in spite of this defect.

5.9 MECHANICAL TRANSFORMERS

The first 'transformer' is the mechanical lever. Here a small force is applied to provide a large force. The transformation is achieved by the fact that the lever pivot is offset from the centre (see Fig. 5.14).

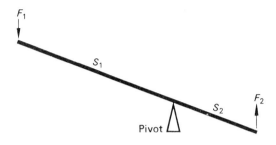

Fig. 5.14

Thus

$$F_1 \times s_1 = \text{input torque}$$
$$= F_2 s_2 \text{ (output torque)}$$

Neglecting any other torque $F_1 s_1 = F_2 s_2$

or $\quad F_1 = F_2 \dfrac{s_2}{s_1}$

Hence if $s_1 > s_2$ then $F_1 < F_2$ and a smaller force (moved through a large distance) produces a much larger force (but moved through a small distance). Ideally the output work = input work. In practice taking into account the mass of the lever itself and the friction at the pivot the output of useful work is always less than the work put in.

The second mechanical transformer is the gear box. Two intermeshing gears with a different number of teeth on each will produce either an increase or a decrease in rotational speed. This is achieved by arranging that the number of teeth on one gear wheel differs from the number of teeth on the other. If one wheel has twice as many teeth as the other then the rotational speed of the output shaft is twice that of the input shaft, assuming that the smaller number of teeth are on the driven shaft (Fig. 5.15). The process is reversible and if the roles of driver and driven are interchanged then the output shaft rotates at half the angular velocity of the input.

Taking a general case, if the number of teeth on the output shaft is a and the number of teeth on the input shaft is na then $na/a = n$ is termed

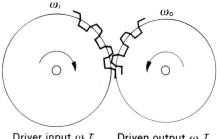

ω_i ω_o

Driver input ω_i T_i **Driven output** ω_o T_o

Fig. 5.15

the gear ratio. The angular velocity of the output shaft is $n \times$ angular velocity of the input or $\omega_o/\omega_i = n$.

The torque transmitted varies inversely in the same ratio if any frictional loss in the gearing is ignored. Thus the output work done $= T_o \times \theta$ where $T_o =$ output torque and $\theta =$ angular movement of the output shaft. The angular movement on the input shaft is θ/n. If there are no losses in the system then $T_i \times (\theta/n) = T_o \times \theta$ where $T_i =$ input torque

$$\frac{T_o}{T_i} = \frac{1}{n} \tag{39}$$

If an increase in torque is necessary then we require more teeth on the output gear wheel than the input.

If a mass of moment of inertia J is coupled to the output shaft the energy stored is $\frac{1}{2}J(\mathrm{d}\theta/\mathrm{d}t)^2$. The input energy (neglecting any losses) must be the same, but the angular movement of the input shaft is one-third that of the output shaft if we assume a gear ratio of 3:

$$\text{Input energy} = \frac{1}{2}J' \left(\frac{1}{3}\frac{\mathrm{d}\theta}{\mathrm{d}t}\right)^2 = \frac{1}{2}\frac{J'}{9}\left(\frac{\mathrm{d}\theta}{\mathrm{d}t}\right)^2 \tag{40}$$

Therefore to maintain the same energy as before $J' = 9J$. The effective moment of inertia on the input with a gear ratio of 3 and a mass of moment of inertia J on the output shaft is 3^2J. In general with a gear ratio of n the 'transformed' moment of inertia is n^2J.

Again the moments of inertia of the gear wheels themselves would modify the 'transformed' moment of inertia slightly.

We can use gearing to translate rotational movement into linear movement by the use of rack and pinion, see Fig. 5.16. The process is reversible if desired. The linear movement can be calculated as follows. Let the effective diameter of the pinion be d m. Then in one revolution a point on the diameter travels a distance πd. Since the rack is in close

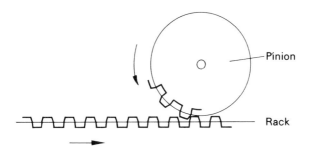

Fig. 5.16

mesh with the pinion the linear movement of the rack must also be πd m.

Therefore an angular velocity on the pinion shaft of n rev/s translates to a linear velocity of πdn m/s.

If the angular velocity ω of the pinion shaft is measured in rad/s then $\omega = 2\pi n$ rev/s. Therefore the linear velocity can be written as πdn or $\pi d\omega / 2\pi = \omega(d/2)$.

A torque T on the shaft results in an input power of $T\omega$ joules. Neglecting losses this must represent the output power.

Thus output work done per second

$$= \text{Force } F \times \text{distance moved per second}$$

$$= F \times \text{velocity}$$

$$= F\pi dn \quad \text{or} \quad F\omega \frac{d}{2}$$

$$\therefore \qquad T\omega = F\omega \frac{d}{2} \quad \text{or} \quad F = \frac{2T}{d} \qquad (41)$$

The input energy to the pinion $= \frac{1}{2} J\omega^2$ if a mass of moment of inertia J is attached to the shaft. The equivalent linear energy $= \frac{1}{2} mu^2 = \frac{1}{2} m(d\omega/2)^2$. Thus (again neglecting losses) $\frac{1}{2} m(d\omega/2)^2 = \frac{1}{2} J\omega^2$ or effective mass

$$m = \frac{J}{(d/2)^2} \qquad (42)$$

Hence a mass

$$m = \frac{J}{(d/2)^2}$$

on the rack has the same inertia effects as a rotational mass of moment of inertia J on the pinion shaft.

73

5.10 ELECTRICAL TRANSFORMERS

The electrical transformer has an input circuit consisting of a number of turns of wire wound around a laminated steel core. If current is allowed to pass through this winding a magnetic flux is produced in the core. A second circuit wound over the first will be linked by this magnetic flux. See Fig. 5.17.

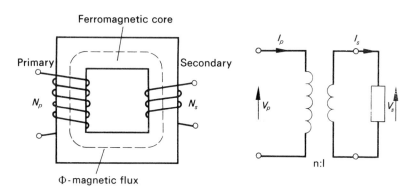

Fig. 5.17

If the number of turns on the input (primary) is N_p and the number of turns on the output (secondary) is N_s then a flux change $d\Phi/dt$ causes a primary emf of $N_p(d\Phi/dt)$ and a secondary emf of $N_s(d\Phi/dt)$.

$$\text{The ratio } \frac{\text{Primary emf}}{\text{Secondary emf}} = \frac{N_p(d\Phi/dt)}{N_s(d\Phi/dt)} = \frac{N_p}{N_s} = n \text{ (the turns ratio)}$$

It follows that

$$\frac{\text{Primary applied voltage } V_p}{\text{Secondary terminal voltage } V_s} = n \text{ (very nearly)} \quad (43)$$

In practice not all the flux generated by the primary links the secondary and this ratio is not exact, but by careful transformer design it can at least be approached.

The secondary circuit may be closed on to a resistor and a current I_s may flow.

$$\text{Thus output power} = V_s \times I_s$$

The input power (neglecting losses) must be the same.

Thus
$$V_p \times I_p = V_s \times I_s$$
or
$$\frac{I_p}{I_s} = \frac{V_s}{V_p} = \frac{1}{n} \tag{44}$$

The ratio $V_s/I_s = R_s$ (the secondary load resistance).
Thus
$$R_s = \frac{V_s}{I_s} = \frac{V_p/n}{nI_p} = \frac{V_p}{n^2 I_p}$$

$$\therefore \quad \frac{V_p}{I_p} = n^2 R_s \text{ or the 'transformed' secondary resistance} = n^2 R_s. \tag{45}$$

We can now make certain comparisons with the mechanical gear box.

Mechanical

Gear box

Gear ratio $n:1$

$\dfrac{\text{Input angular velocity}}{\text{Output angular velocity}} = \dfrac{1}{n}$

$\dfrac{\text{Input torque}}{\text{Output torque}} = n$

'Transformed' moment of inertia from output to input $= n^2 J$

Electrical

Transformer

Turns ratio $n:1$

$\dfrac{\text{Input current}}{\text{Output current}} = \dfrac{1}{n}$

$\dfrac{\text{Input voltage}}{\text{Output voltage}} = n$

'Transformed' resistance from output to input $= n^2 R$

The last comparison must be commented on. $n^2 R$ is a transformed resistance or dissipative term in the electrical case. $n^2 J$ is a transformed non-dissipative element, i.e. a reactance. The same principle applies however whatever form of element dissipative or non-dissipative, mechanical or electrical we are using.

5.11 SUMMARY

Transducers are devices which translate energy from one form to another. They often translate non-electrical quantities into electrical. They can be used to measure, for example, temperature, pressure, rotational speed and acceleration.

Transformers are devices which change the magnitude of a parameter within a given system. The electrical transformer changes voltages or currents; the mechanical transformer (the gear box) changes torque.

Elementary Engineering Systems

Sensors and transmitters are used to sense and transmit information (often of the position of an object). Actuators are devices capable of producing large power outputs from relatively small input signals.

QUESTIONS

5.1 Four identical strain gauges are mounted on a beam; two are in tension and two in compression when the beam is deflected. The gauges are connected in the form of a bridge, to give the maximum output. Show how they are connected. The bridge supply is 5 V and the gauge factor is 2·2. If the strain in the beam is 500×10^{-6}, find the output.

5.2 Draw a diagram of a typical accelerometer and explain how it operates.

5.3 In the diagram shown the polar mass moment of inertia of the drum and gear wheel A is 8 kg m², while that for the flywheel and gearwheel B is 15 kg m². Find the linear acceleration of the 5 kg mass.
Produce also an equivalent electrical circuit.

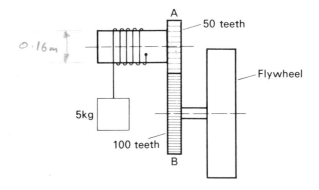

5.4 The angular acceleration of a rotating device is measured by using a differentiating circuit consisting of a resistor and a capacitor connected across a tachogenerator, which gives an output of 0·15 V per rad/s. If $R = 10$ kΩ and $C = 0·05$ μF find the voltage available from the differentiating circuit if the rotating device is subjected to an angular acceleration of 25 rev/s². The errors introduced by the differentiating may be neglected.

5.5 Eddy current damping or braking is frequently used in control systems. It consists of a thin copper disc which passes between the poles of a magnet. Show that the braking torque produced by this method is directly proportional to the angular velocity of the disc.

ANSWERS

5.1 5·5 mV   **5.3** 0·029 m/s²   **5.4** 11·78 mV

76

Chapter 6
Amplifiers

6.1 INTRODUCTION

An amplifier is a device for producing a magnified replica of an input variable or signal. The amplifier is said to be linear if the output retains the same characteristics as the input.

A microscope might be termed an image amplifier. A system of lenses produces a magnification of an object so that the object appears very much larger than it really is. It is necessary to supply additional illumination in order to see the object, that is, additional light energy is needed to make the microscope operate satisfactorily.

An electronic amplifier is one in which an input electrical signal of current or voltage (or both) is magnified. It is necessary to supply additional electrical energy to the system (from batteries or rectified mains supply) in order that the system will function.

Since the main purpose of the amplifier is to magnify, the overall efficiency is usually not of major importance and high efficiency is often traded for accurate reproduction of the input variable.

Fluid amplifiers exist, where small variations of input pressure are reproduced as large variations of output pressure. In each case the amplifier does not change the energy system employed.

Amplifiers are not transducers. Neither are amplifiers transformers, in that they are capable of power or energy magnification of the signal. To make them function it is necessary to supply energy from an additional source and the application of an input signal effectively diverts some of this energy to the output. If the system is essentially a power or energy amplifier, then as much as possible of this additional energy must be transferred to the output, but often the variable which is to be amplified is *not* power or energy. The amplification is referred to as 'Gain'.

6.2 DISTORTION

Amplifiers usually introduce unwanted distortion where the output is not a faithful reproduction of the input. The distortion can be listed under three main headings:

1. Frequency distortion (or amplitude distortion)
2. Phase distortion
3. Harmonic distortion

1. The first is due to the fact that amplifiers do not amplify signals at all frequencies to the same extent. Some amplifiers are designed to handle a wide range of frequencies, others a small range. The term 'bandwidth' is employed to define the range of frequencies that an amplifier will handle. Wide bandwidth infers a large frequency range. In general the wider the bandwidth of a given amplifier the smaller the gain.

Ideally an amplifier dealing with signals from say 0·1 Hz to 10 kHz should have a uniform gain over this bandwidth. In practice there is often a tendency for the gain to fall off at the top and bottom ends of the frequency range (Fig. 6.1). Consequently the amplification produced of signals at the extreme ends of the frequency range is often less than that in the middle of the frequency band.

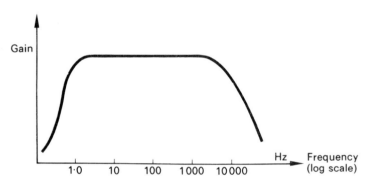

Fig. 6.1

2. Amplifiers usually produce phase shifts, that is a phase or time delay may occur between output and input signal. The phase shift is often non-uniform over the frequency range and certain frequencies will suffer bigger time delays than others. This is phase distortion.

3. Due to non-linearities in the amplifier frequencies which do not exist in the input appear in the output. This is harmonic distortion. It may be worthwhile at this point to show how these additional frequencies are produced.

Let the output pressure p_o of some form of amplifier be related to the input pressure p_i by the equation $p_o = A p_i$ where A is the amplification or gain.

If A is 50 then the output ρ_o is 50 times the input ρ_i. If ρ_i is a sine-wave, $\rho_i = \rho \sin \omega t$, then $\rho_o = A\rho \sin \omega t$ and no distortion occurs. This is a linear case. The relationship between ρ_o and ρ_i may not however be linear. If the equation relating ρ_o and ρ_i is

$$\rho_o = A\rho_i + B\rho_i^2 \text{ the last term } B\rho_i^2 \text{ produces distortion.}$$

Thus if $\rho_i = \rho \sin \omega t$ as before

$$\rho_o = A\rho \sin \omega t + B\rho^2 \sin^2 \omega t$$

Now $\sin^2 \omega t$ can be expressed as $\frac{1}{2}(1 - \cos 2\omega t)$

$$\therefore \qquad \rho_o = A\rho \sin \omega t + \frac{B\rho^2}{2} - \frac{B\rho^2}{2} \cos 2\omega t \qquad (46)$$

The term $\frac{1}{2}B\rho^2$ is a zero frequency term which was certainly not present in the input. The term $\frac{1}{2}B\rho^2 \cos 2\omega t$ is a function with a frequency twice that of the input frequency. This component is termed a harmonic of the input and $\frac{1}{2}B\rho^2 \cos 2\omega t$ is a harmonic distortion component. Being twice the input frequency it is termed second harmonic distortion.

The inputs and outputs to four amplifiers are shown in Fig. 6.2. Only the first is regarded as a truly 'linear' amplifier. In practice we have to put up with a certain amount of distortion but we design the amplifier so as to be as free of these defects as possible.

6.3 AMPLIFIER DESIGN

It is not the intention of the text to deal with the way amplifiers are designed but merely to point out how amplifiers must be specified to fit into a system. Electronic amplifiers can be bought 'off the shelf' so that often it is necessary to specify fully what is required.

Obviously we need to specify whether the output should be essentially voltage, current or power and we need to know the form of the input variable. We need to specify the gain and bandwidth required and we therefore wish to know the distortion defects of the amplifier. Size and weight may also be important parameters – although usually electronic amplifiers are very small and very light. We need to know the electrical supplies required by the device (normally referred to as high tension or h.t. supplies). We must also be aware of the environment in which the amplifier is to be employed: is it at a high temperature?, a chemically active (i.e. corrosive) atmosphere?, is it likely to be subjected to a great deal of vibration? We may wish to know how stable the gain is and how reliable the amplifier is. Amplifiers are often cheap enough to discard when shown to be faulty but if they are

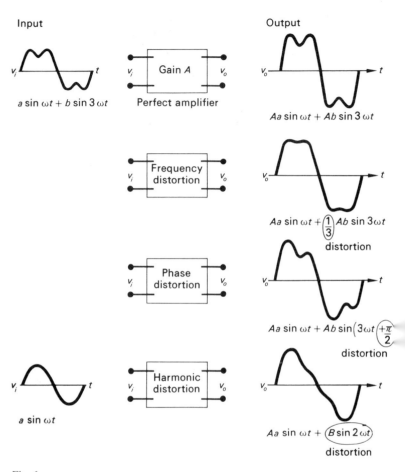

Fig. 6.2

situated in an inaccessible location the overall cost of replacement may be many times the cost of the amplifier itself.

There are however two further parameters which must be specified. They are the input and output impedances of the device. A low input is generally needed for a current amplifier – a high input impedance for voltage. Similarly on the output side a high output impedance is desirable for a current output and a low output impedance for voltage. This coupling between the input of one electrical system and the output of another is referred to as matching. A very bad mismatch may result in an effective reduction (attenuation) of the signal rather than in

4. A hybrid analogue system

amplification. Matching is not a highly critical parameter and fairly wide variations in input and output impedances may have only minor effects except in the case of power transfer. Maximum power transfer can occur only when the output impedance of the amplifier is the same as the impedance to which it is connected. If a serious mismatch exists then an output transformer may have to be used. For example the load device to which the output of an amplifier is to be connected may have a resistance of 50 Ω and the amplifier may have an output resistance of 10 000 Ω. A serious mismatch would exist connecting the amplifier to the device directly and maximum power available from the amplifier could not be transferred to the load. We have shown already however that a resistance R on the secondary of a step-down transformer of turns ratio n appears as an effective resistance n^2R on the primary side. In our example $R = 50$ Ω and for matching purposes n^2R must equal 10 000 Ω.

hence $n^2 50 = 10\ 000$ $n^2 = 200$ $n = 14 \cdot 1$

We therefore need a matching transformer of turns ratio about 14 : 1 between the device and the amplifier. Note that this method of matching can only be used under alternating current conditions since transformers are essentially a.c. devices. Table 6.1 gives a list of amplifier specifications. Where the amplifier is designed to produce an output voltage from an input voltage (or an output current from an input current) the term 'gain' is used to give the ratio of output to input. Where the output is essentially current for an input voltage the term 'gain' is meaningless and the expression 'transfer function' is more applicable. A transfer function of 500 mA/V means that the output current increases by 500 mA for every input volt. Usually a given load resistor is specified.

Design considerations	Manufacturer's specifications
Voltage, current or power amplification	Type
Frequency range	Bandwidth
Gain needed	Nominal gain
Input impedence	Nominal input impedance
Output impedance	Nominal output impedance
Adverse environment	Maximum ambient temperature
Size	Size
Weight	Weight
Reliability	
Cost	Cost
H.t. supplies	H.t. supplies

Table 6.1

Electronic Amplifiers

6.4 THE OPERATIONAL AMPLIFIER

An operational amplifier is a high gain amplifier operating over a range of frequencies from zero to a few hundred hertz. It is used to perform certain mathematical operations.

It consists essentially of the amplifier itself (which might be highly complex) and certain additional external components which enable it to operate according to certain mathematical relationships. The input impedance to the amplifier is high and the output impedance (the internal impedance) is low. A series input resistance R_i is included in one arm of the input. A second resistance R_f is connected as shown in Fig. 6.3 between input and output.

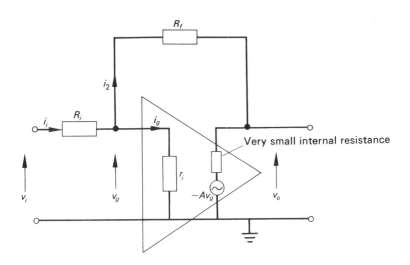

Very small internal resistance

Usually redrawn as:

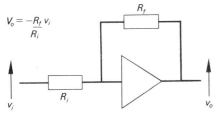

All voltages with reference to earth potential

Fig. 6.3

Let v_i = input voltage to the circuit

v_g = input voltage to the amplifier itself

v_o = output voltage of the amplifier

$-A$ = gain (the negative sign shows that v_o is in antiphase to v_i)

r_i = input resistance of the amplifier

We will assume no distortion.

Now $i_1 = \dfrac{v_i - v_g}{R_i}$ $i_2 = \dfrac{v_g - v_o}{R_f}$ $i_g = \dfrac{v_g}{r_i}$ $\dfrac{v_o}{v_g} = -A$

Thus by Kirchhoff's law at the input $i_1 = i_2 + i_g$.

Hence $\dfrac{v_i - v_g}{R_i} = \dfrac{v_g - v_o}{R_f} + \dfrac{v_g}{r_i}$

But $v_g = -\dfrac{v_o}{A}$

$\therefore$ $\dfrac{v_i + v_o/A}{R_i} = \dfrac{-v_o/A - v_o}{R_f} - \dfrac{v_o/A}{r_i}$

If A is very large – say 10^6 – and v_o and v_i are of roughly the same magnitude and r_i is at least as large as R_i and R_f we can neglect all terms containing v_o/A.

Hence

$$\dfrac{v_i}{R_i} = \dfrac{-v_o}{R_f}$$

or

$$v_o = -v_i \dfrac{R_f}{R_i} \tag{47}$$

This result is interesting in that the magnitude of the output voltage v_o is independent of the gain of the amplifier (providing A is large enough) and depends only upon the ratio of the two resistors employed.

If $R_i = R_f$ then v_o is v_i reversed in phase.

If $R_i = \frac{1}{10}R_f$ then v_o is $10 \times v_i$ but reversed in phase.

The gain can be adjusted to what we wish by changing R_i or R_f. Typical values of R_i and R_f lie in the range 100 kΩ to 1 MΩ. It should be noted that v_g, the voltage at the input to the amplifier itself, is extremely small and can be regarded as effectively zero.

6.6 PHASE REVERSAL

If the sole object is to change the polarity of a voltage then we can simply employ an operational amplifier in which $R_i = R_f$. Then $v_o = -v_i$.

6.7 SUMMATION OF TWO VOLTAGES

Let two resistors R_1 and R_2 be employed as shown in Fig. 6.4. Let v_1 and v_2 be applied to each of these.

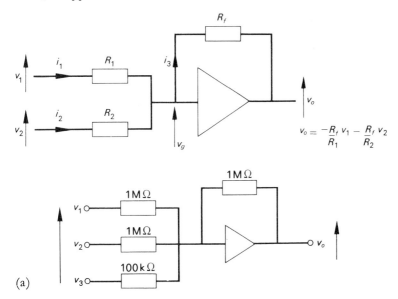

(a)

Fig. 6.4

Hence $i_1 = v_1/R_1$ and $i_2 = v_2/R_2$ if v_g is regarded as effectively zero.

$$i_3 = \frac{v_g - v_o}{R_f} = -\frac{v_o}{R_f} \text{ neglecting } v_g.$$

Hence

$$i_1 + i_2 = i_3$$

$$\frac{v_1}{R_1} + \frac{v_2}{R_2} = -\frac{v_o}{R_f} \tag{48}$$

and if $R_1 = R_2 = R_f$ then $v_o = -(v_1 + v_2)$ and summation of two voltages occurs, again with a phase reversal.

Since the input and output earth connections are common we usually omit these from the diagram. Fig. 6.4(a) shows a summing circuit:

$$v_o = -(v_1 + v_2 + 10v_3)$$

The voltages v_1, etc., may of course be changing providing the frequency at which they are changing is within the bandwidth of the amplifier.

6.8   SUBTRACTING TWO VOLTAGES

If v_2 is to be subtracted from v_1 then we wish to change the sign of v_2 and add to v_1, i.e. $v_1 - v_2 = v_1 + (-v_2)$. Two operational amplifiers are necessary, the first used as a phase reversal the second as an adder. See Fig. 6.5.

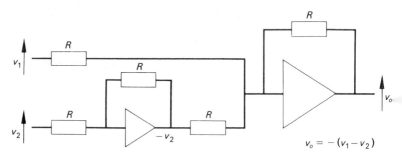

Fig. 6.5

6.9   INTEGRATION

If the resistor R_f is replaced by a capacitor C then the circuit behaves as an integrator (Fig. 6.6).

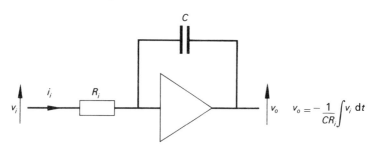

Fig. 6.6

Thus $i_1 = v_i/R_i$ as before neglecting v_g.
The voltage across C is $v_g - v_0$ which is $-v_0$ again neglecting v_g.

∴   Charge q on $C = -Cv_0$

  Charge $= i \times t$, therefore current $= \dfrac{\mathrm{d}q}{\mathrm{d}t}$

5.   A small electronic computer

$$\therefore \quad i_2 = \frac{-\mathrm{d}(Cv_0)}{\mathrm{d}t} = -C\frac{\mathrm{d}v_0}{\mathrm{d}t}$$

But $i_1 = i_2$

Hence $\dfrac{v_i}{R_i} = -C\dfrac{\mathrm{d}v_0}{\mathrm{d}t}$

or $\mathrm{d}v_0 = -\dfrac{1}{C}\dfrac{v_i}{R_i}\,\mathrm{d}t$

Integrating

$$v_0 = -\frac{1}{CR_i}\int v_i\,\mathrm{d}t \tag{49}$$

The output voltage v_0 is proportional to the integral with respect to time of the input voltage v_i (again with a phase reversal)

If CR_i is made equal to unity

$$v_0 = -\int v_i\,\mathrm{d}t$$

R_i is typically 1 MΩ hence C must be 1 μF for the time constant CR_i to be unity.

6.10 DIFFERENTIATION

If the positions of R_i and C are interchanged (Fig. 6.7) then a differentiating circuit results

$$v_0 = -R_iC\frac{\mathrm{d}v_i}{\mathrm{d}t} \tag{50}$$

Such a circuit should be avoided if possible as it passes unwanted
signals and the circuit tends to be unstable.

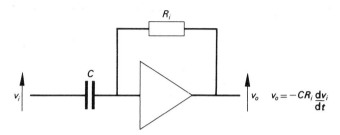

$$v_o = -CR_i \frac{dv_i}{dt}$$

Fig. 6.7

6.11   THE ELECTRONIC ANALOGUE COMPUTER

Operational amplifiers can be connected together so that the circuit
behaves as a mathematical model. Take for example the differential
equation

$$F = a\frac{ds}{dt} + ks$$

This equation represents a spring/damping system subjected to a
force F.

To make the work easier initially let us assume that the coefficients
a and k are both unity.

Then $F = \dfrac{ds}{dt} + s$

rearranging

$$-s = \frac{ds}{dt} - F$$

A circuit which will perform the same function is shown in Fig. 6.8.
At point f we have a voltage representing ds/dt. The first amplifier
circuit (1) integrates and produces a phase reversal. Therefore we have
at point c a voltage representing $-s$. At point b we have the force F.
We have to represent F by an applied voltage. Amplifier circuit (2)
merely reverses F, i.e. adds a minus sign. Amplifier circuit (3) now
adds $-F$ and ds/dt and produces an output $-ds/dt + F$. Amplifier
circuit (4) effects a further phase reversal so that at point d we have
$ds/dt - F$.

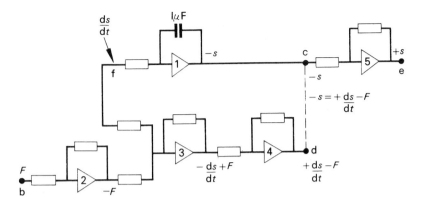

Fig. 6.8

If we now link points d and c we equate $-s$ and $ds/dt - F$, i.e. we have $-s = ds/dt - F$ which is the original equation. If we wish now to see how s varies with time we need a further amplifier circuit 5 to reverse $-s$.

If we need to know the way the system responds to a step input F we apply a sudden positive voltage to point b and observe the voltage at e. We can easily dispense with amplifiers 2 and 5 if (i) we apply a negative voltage for F and (ii) we appreciate that the voltage we observe at c is $-s$ and not $+s$. (It is actually possible to remove amplifiers 3 and 4 but this is left to the reader to sort out.)

If the coefficients a and k have to be taken into account then we use the equation

$$\frac{F}{a} = \frac{ds}{dt} + \frac{k}{a} s$$

Rearranging $\quad -\dfrac{k}{a} s = \dfrac{ds}{dt} - \dfrac{F}{a}$

If $a > 1\cdot0$ and $k/a < 1\cdot0$ then we can use the circuit shown in Fig. 6.9. Potential dividing circuits enable a fraction of $-F$ and a suitable fraction of $-s$ to be applied to the rest of the circuit. If k/a had a value between 1 and 10 the input resistor on the integrating amplifier could be changed to 100 kΩ. This would give an output $-10s$ and the potential dividing circuit would provide the necessary scaling factor. The diagram showing the interconnection of the various operational amplifiers is called a flow diagram. It is necessary to give a voltage scale, i.e. in this case to state how many newtons are represented by 1 volt.

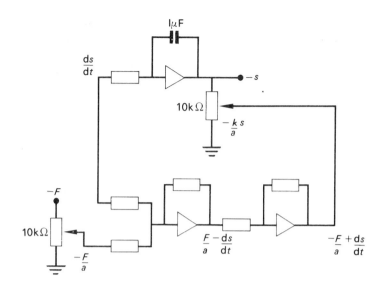

Fig. 6.9

6.12 A SECOND-ORDER EQUATION – PRODUCTION OF S.H.M.

The equation of S.H.M. can be written thus for an electrical oscillatory system:

$$L\frac{d^2q}{dt^2}+\frac{1}{C}q=0$$

Thus $\dfrac{d^2q}{dt^2}=-\dfrac{1q}{LC}$

Three amplifiers are necessary to simulate this system (Fig. 6.10). At a we have d^2q/dt^2, at b $-dq/dt$ and at c $+q$.

If the input resistor to the third amplifier is 100 kΩ the output is $-10q$ and suitable scaling is effected by the potential dividing circuit. Joining d and a we have our second-order system. In order to generate the oscillations it is necessary to disturb the system. One way of doing this is to have an initial charge on one of the capacitors and arrange for this to be discharged when the circuit is completed.

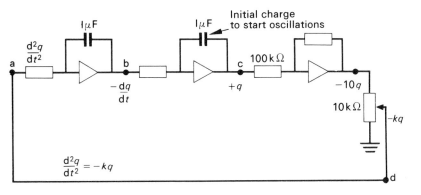

Fig. 6.10

6.13   A SECOND-ORDER EQUATION WITH DAMPING

An equation representing a torsional second-order system is

$$J\frac{d^2\theta}{dt^2} + C\frac{d\theta}{dt} + K\theta = T_0$$

Dividing by J

$$\frac{d^2\theta}{dt^2} + \frac{C}{J}\frac{d\theta}{dt} + \frac{K}{J}\theta = \frac{T_0}{J}$$

and rearranging

$$\frac{d^2\theta}{dt^2} = -\frac{C}{J}\frac{d\theta}{dt} - \frac{K}{J}\theta + \frac{T_0}{J}$$

The analogue computer circuit which will achieve this result is shown in Fig. 6.11. Providing C/J, K/J and $1/J$ are all less than unity these coefficients can be obtained by the use of potential dividing circuits.

T_0/J is applied by external means. If this is a step function then it is achieved by suddenly switching on a direct voltage. If a sinewave forcing function is required then the input voltage representing $T_0/J \sin \omega t$ is obtained from a suitable low frequency oscillator.

6.14   ANCILLARY EQUIPMENT

Analogue computers require a certain amount of additional equipment in order that the result of any computation may be recorded. An oscilloscope is a useful method of displaying the output quantity but

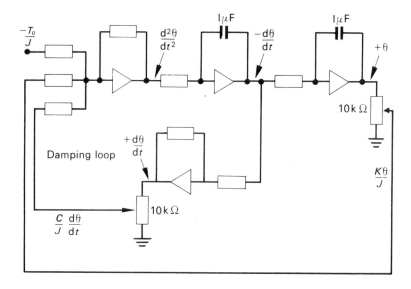

Fig. 6.11

since the frequencies involved in computation are usually low an XY plotter is used. This is a device in which two input voltages control the position of a recording pen in two axes. The Y displacement is the signal to be observed and the X voltage is usually a linearly increasing quantity – similar to the time base of a cathode-ray oscilloscope. Other equipment is usually limited to a low frequency oscillator. The feedback components (capacitors and resistors) are usually part of the computer and can be switched into the circuit as required. The connections between the amplifiers are made by flexible 'plug in' leads.

Electronic analogue computers are versatile and useful tools to provide working models of systems of all types. Other attempts have been made to produce fluid analogue computers or in one case a computer based on mechanical movement of steel tapes. They suffer from their difficulty of 'programming' and are consequently little used.

6.15 SUMMARY

Amplifiers form important parts of systems. It is usually necessary to know their input and output characteristics as well as the gain. Amplifiers are liable to produce distortion. Operational amplifiers perform mathematical operations and can be linked together to produce models of systems. Response of the analogue to forcing functions will give the response of the system under examination.

QUESTIONS

6.1 The gain characteristic of an amplifier is given by $\theta_o = 60\theta_i + 5\theta_i^2$ where θ_o = output, θ_i = input. If θ_i is a sinewave $3 \sin 400t$ give an expression for the output and find the magnitude of the second harmonic component.

6.2 A power amplifier is capable of delivering 500 W at a frequency of 100 Hz. The output resistance of the amplifier is 250 Ω and the load resistance is 20 Ω. What ratio of turns is required from a matching transformer for maximum power outupt. If the transformer is 90 per cent efficient and the amplifier is fully loaded what is the voltage across the load resistor?

6.3

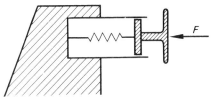

The diagram represents a railway buffer stop. The dashpot has a damping coefficient of 400 Ns/m and the spring has a compression stiffness of 300 N/m. If the force F is suddenly applied and maintained constant at 1 500 N produce a flow diagram for an analogue computer to give the displacement of the buffer.

The operational amplifiers have input resistors of values 1 MΩ and 100 kΩ. The feedback path of each amplifier can be arranged to be either a 1 μF capacitor or a 1 MΩ resistor. Using a suitable voltage scale draw a flow diagram to represent this system and insert the values of components. A 50 V d.c. reference voltage is available (either positive or negative polarity) and two 50 kΩ potentiometers are supplied. Indicate the setting on any potentiometer.

6.4

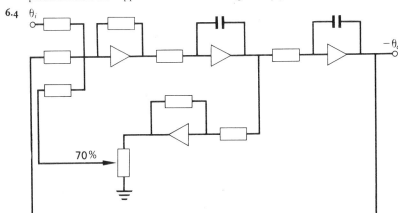

All fixed resistors 1 MΩ All capacitors 1 μF

What is the mathematical relationship between θ_o and θ_i?

6.5 An operational amplifier has a feedback resistance of 250 kΩ. It has three inputs A, B and C with input resistors 250 kΩ, 100 kΩ and 500 kΩ respectively. Input voltages of +5 V, −10 V and +1·5 V are applied to the three inputs respectively. What is the output voltage?

6.6 If the amplifier used in the previous question has the following input voltages applied give the equation of the output voltage

        A: 10 sin 500 *t*
        B: 5 sin 500 *t*
        C: 20 cos 500 *t*

6.7 An integrating amplifier uses $C = 0.5$ μF and $R = 0.5$ MΩ. It saturates when the output reaches 80 V. If an input voltage of −10 V is applied what time must elapse before saturation?

6.8 An operational amplifier has a feedback capacitance of 1 μF and two inputs each with 500 kΩ resistors. Pulses of voltage are applied to each input, the pulses are each of magnitude +5 V and duration 0·5 s but there is a time displacement of 0·2 s between them. Sketch to scale the output voltage.

6.9 The following waveforms are applied in turn to an integrating amplifier $C = 1$ μF, $R = 0.5$ MΩ. Sketch to scale the output voltages.

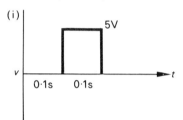

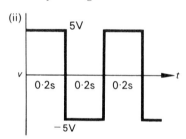

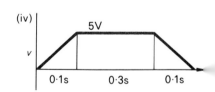

ANSWERS

Chapter 7
Control Systems

7.1 INTRODUCTION

We live in a world where machines do much of the work previously undertaken by man. The machines lift, transport and process materials and the human operator is merely there to control them. It is sometimes even possible to dispense with the services of the human operator – we then have a system of automatic control.

Such a system has certain advantages over one involving a human operator. It does not suffer from fatigue and can therefore work continuously, if necessary, over long periods. It is consistent, unlike the man controlled system which is liable to human error and judgement. It can operate in environments in which a human being could not safely work, e.g. a radioactive atmosphere. It can be made to operate very quickly and can therefore have a much faster response time than the human system.

But automatic control systems are expensive, often sophisticated and may occasionally go wrong.

7.2 THE CASE FOR AN AUTOMATIC SYSTEM

Let us take one or two simple examples. Firstly, let us consider a car ramp or lift for raising cars above the ground to enable an inspection to be made of the underside of the body. The car ramp could be raised by a human operator. It would naturally be necessary to have some mechanical means whereby the available human effort (which is small) could be transformed. A suitable worm gearing and racket mechanism could be employed but the human effort would have to be spread over a considerable period to enable the car to be lifted.

The car might have a mass of about 1 000 kg and if it is to be lifted 1·5 m then the mechanical work needed is 1 000 × 9·81 × 1·5 Nm or joules. Allowing 75 per cent efficiency for the gearing the input work required is

$$\frac{100}{75} \times 1\,000 \times 9 \cdot 81 \times 1 \cdot 5 = 19\,620 \text{ J.}$$

If man's effort is restricted to 50 J/s the time to lift the car is

19 620/50 = 392 s or 6·5 min. This is based on the assumption that the person operating the ramp works continuously for this period with a steady work output of 50 J/s. Allowing for a 'breather' and physical fatigue a realistic estimate for the time required to raise the car is of the order of 10 min.

A small electric motor with ten times the output power could achieve the same result in 39 s, removing completely the human fatigue factor. The only human effort needed is that required to switch the motor ON and OFF.

The human operator is there to make two decisions (i) when to switch ON and (ii) when to switch OFF.

7.3   A SIMPLE AUTOMATIC CONTROL SYSTEM

Indeed it is possible by means of a suitable switch which is operated by the ramp itself to dispense with the services of the human being after switching ON. The motor would have to be switched OFF automatically when the ramp had reached the required height.

The result is:

(a) A human operator is only necessary to start the operation.
(b) The operation is completed speedily.
(c) The operator does not suffer from fatigue.

We now have a very simple automatic control system which is shown diagrammatically in Fig. 7.1.

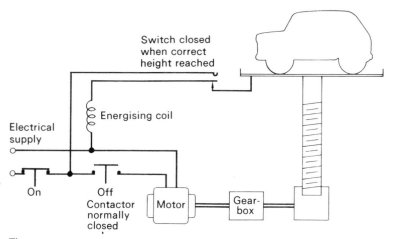

Fig. 7.1

The link between the output (vertical height of the ramp) and the input electrical energy is via an electromagnetic relay operated by two contactors – one for ON the other for OFF. This link between the ramp and the OFF contactor is an essential part of the automatic system. The link is termed a feedback path and must provide the information telling when the ramp is in the required position. It must hold the OFF contactor closed all the while there is any difference between the actual position and the required position of the ramp. It must open the contactor as soon as this difference falls to zero. The difference between the required and actual levels is called the ERROR. When the error is zero the required level has been reached. Such a system is called a closed-loop system. If no feedback link is provided the system is said to be open loop and would require the services of a human operator to turn the motor off. This type of closed-loop system is an ON/OFF system, i.e. the driving motor is either ON or OFF.

7.4 A THERMAL AUTOMATIC CONTROL SYSTEM

Another example of an ON/OFF closed-loop system occurs in the thermostat and electric immersion heater found in a typical hot-water system. The electricity supply is connected via a switch to the heating element in the hot-water tank. A temperature sensitive device usually acting on the bimetallic strip principle opens the switch when the water has reached the required temperature and closes the switch when the temperature falls. See Fig. 7.2. The temperature of the water is main-

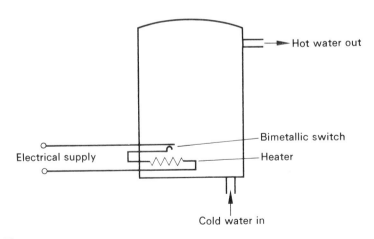

Hot water out

Bimetallic switch

Electrical supply

Heater

Cold water in

Fig. 7.2

tained constant against heat losses, by the heater being periodically switched on automatically by the thermostat (the heat sensitive bimetal strip). There will be a slight fall in temperature before the switch operates because of a 'dead zone' – i.e. the difference between the required temperature and the actual temperature (the error) must exceed a certain value before the switch recloses. For example if the required temperature is 80°C and the thermostat is set to this temperature a fall of perhaps 2°C is needed before the heater is switched on again once the temperature has been reached. The switching period for maintaining constant (or nearly constant) temperature is increased in length if the tank is thermally 'lagged' to reduce heat losses.

When hot water is drawn off cold water enters the tank and the temperature falls rapidly. The heater is then switched on automatically and the water heats up again. There is a time lag before the required temperature is reached since some time must elapse before sufficient energy generated by the electric heater is transferred to the water. The time lag depends upon the amount of cold water entering the tank and its temperature and also the power output from the immersion heater.

Figure 7.3 shows three graphs indicating the variation of temperature with time.

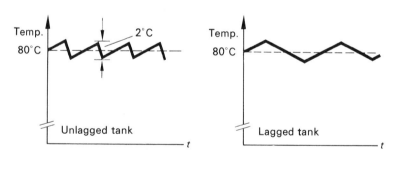

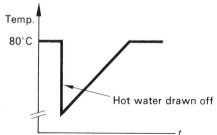

Fig. 7.3

The required temperature is called the 'set point' and the difference between 'set point' temperature and actual temperature is the error.

7.5 A PROPORTIONAL CONTROL SYSTEM

Let us now consider another domestic situation – the lavatory cistern, Fig. 7.4. When the chain or lever is operated the cistern empties

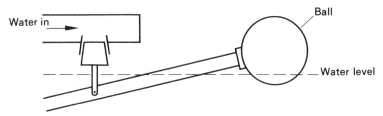

Fig. 7.4

rapidly. The cistern is then refilled through a ball valve. As the level of water rises the ball is lifted, partially closing the valve and the closure increases progressively as the water level rises, finally closing completely when the required amount of water is in the cistern. The flow of water decreases as the difference between the required water level and the actual level decreases.

This difference is the error and the inflow of water is directly dependent upon the error. This is a proportional system, i.e. inflow is proportional to error.

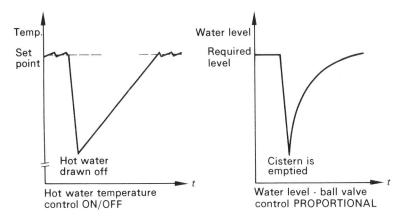

Fig. 7.5

We can now compare the response of the two systems ON/OFF and proportional. See Fig. 7.5.

The second example shows the response curve to be exponential in shape – typical of 'proportional' systems. The first example shows a linear response but a continuous switching action occurs when the set point is reached. There would obviously be many control systems when the ON/OFF type of control would be quite unacceptable, e.g. a crane.

7.6 SOME GENERALISATIONS

Let us look at these three examples of a closed-loop system and see what they have in common.

1. Each needs some form of 'sensing' element to detect the error.
2. Each controls some variable (position, temperature, level).
3. Each has a feedback loop.
4. Each has some energy input (electric motor, electricity supply, water pressure).
5. Each has some desired result in the output (the car raised to the right position, the water temperature raised to the right value, the water level raised the correct amount).

They can all be represented by a block diagram as in Fig. 7.6. Here is shown a symbol indicating some means of comparing the actual result

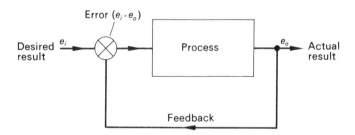

Fig. 7.6

and the desired result (error sensing) which controls the input to the system, the system itself, an input e_i and an output e_o. e_i and e_o can indicate position level, voltage temperature or any system variable.

The error $(e_i - e_o)$ controls the input to the process making it change towards the desired value. When that is reached the error $e_i - e_o$ falls to zero and no further change occurs in the process. In order to reduce the error the feed-back quantity e_o must be subtracted from e_i, i.e. e_o must have a negative sign associated with it. Because of this the feed-back described is termed NEGATIVE feedback.

In the case of the car ramp and the temperature control the feedback was discontinuous – either OFF or ON. With the lavatory cistern the amount of feedback depended upon the error.

7·7   A REMOTE POSITION CONTROL SYSTEM

We will now consider a more sophisticated situation, the control of the horizontal angular position of a large steerable radio telescope.

It is obvious that some assistance is necessary to move the massive structure which may weigh many tonnes. (1 tonne = 1 000 kg.) We must therefore use a motor to assist us. A motor used in this way is called a 'servo motor'. We can use a d.c. motor with a constant field excitation so that the armature current provides the method of speed and torque control. It will be necessary to include a suitable gearbox, with a considerable gear ratio between the motor and the telescope, i.e. the angular velocity of the output shaft must be much smaller than the velocity of the input. The system is represented diagrammatically by Fig. 7.7.

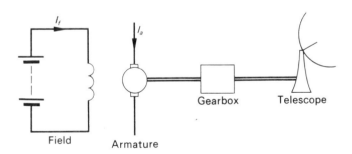

Fig. 7·7

The speed and direction in which the telescope moves is controlled by the magnitude and direction of the current I_a through the armature. As shown the system is open loop – a human operator has to control I_a until the telescope is in the desired position.

The system can be adapted to closed-loop operation by the addition of two potentiometers. See Fig. 7.8.

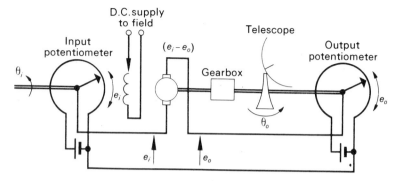

Fig. 7.8

The input potentiometer arm is moved by hand to the desired position and a voltage e_i appears between the end of the potentiometer and the moving contact; e_i is proportional to $\theta_i (e_i = r\theta_i)$. The output potentiometer arm is attached to the steerable telescope so that the voltage e_o between the contact and one end of the potentiometer is proportional to the angle turned through θ_o ($e_o = r\theta_o$). The voltage applied to the armature is therefore $e_i - e_o$ (the error), and armature current flows all the while $e_i \neq e_o$. We have a proportional system because the current flowing through the armature is directly dependent upon the error and this current falls to zero when $e_i - e_o = 0$, that is the desired position has been reached. This is again an example of negative feedback.

There is a major drawback to this arrangement. The armature current required to move the telescope initially may be quite large and if the error is small the voltage $e_i - e_o$ may not be sufficient to drive the necessary current through the armature and no movement occurs. We have a 'dead zone' again. It is therefore necessary to amplify the error to provide the required current for small errors. Rearranging the circuit slightly and introducing an amplifier we have Fig. 7.9.

It is possible now to put in some other parameters and consider the whole closed-loop system. Firstly, there is obviously a considerable inertia in the system. Let us assume that the moment of inertia of the telescope is J. Let the angle of the output be θ_o.

The inertia torque to be overcome is $J(d^2\theta_o/dt^2)$ ($J \times$ angular accelera-tion). The system has a certain resistance to angular movement produced by friction at the bearings, air friction at the telescope and the driving motor, friction in the gear box. Such friction is called 'viscous friction' and is usually taken to be directly proportional to the angular velocity, that is

$$\text{viscous friction} = C\frac{d\theta_o}{dt}$$

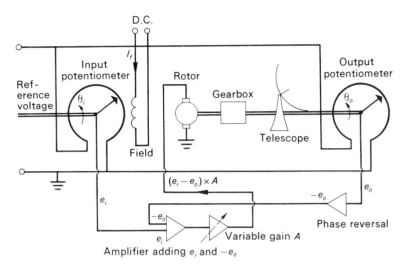

Fig. 7.9

C must represent the total viscous friction of the whole system. The torque produced is dependent upon the error voltage $e_i - e_o$ which in turn is dependent upon the setting of the input and output potentiometers. That is error torque $\propto r\theta_i - r\theta_o$

where $r =$ potentiometer constant in V/rad.

Thus error torque $= K(\theta_i - \theta_o)$

The system equation becomes

$$J\frac{d^2\theta_o}{dt^2} + C\frac{d\theta_o}{dt} = K(\theta_i - \theta_o)$$

or

$$K\theta_i = J\frac{d^2\theta_o}{dt^2} + C\frac{d\theta_o}{dt} + K\theta_o \tag{51}$$

If we wish to take the gear box ratio and the amplifier gain into account then the best method is to relate all the torques in the system to one shaft, either the output shaft or the motor shaft. Let the inertia of the whole system referred to the motor shaft be J_m and the viscous friction coefficient of the whole system referred to the motor shaft be C_m. The input to the amplifier is $r(\theta_i - \theta_o)$ and the gain A produces an output $Ar(\theta_i - \theta_o)$. Thus the torque at the motor shaft produced by this output is $MAr(\theta_i - \theta_o)$, where M relates the output motor torque to armature (or rotor) current. M is the motor transfer function. The

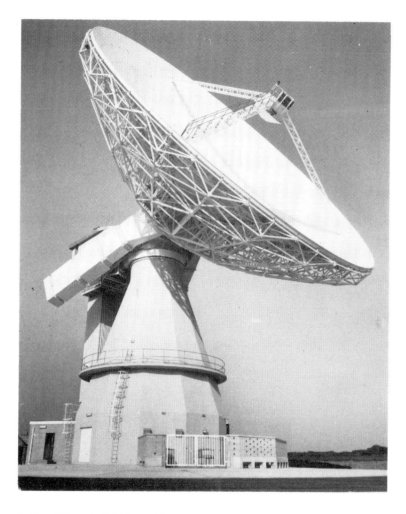

6. Goonhilly 3 Aerial, Cornwall

system equation now becomes

$$J_m \frac{\mathrm{d}^2\theta_m}{\mathrm{d}t^2} + C_m \frac{\mathrm{d}\theta_m}{\mathrm{d}t} - MAr(\theta_i - \theta_o),$$

θ_m being the angular movement of the motor shaft.

Now $\theta_m = n\theta_o$ where $n =$ gear ratio and $\theta_o =$ angular movement of the output shaft.

The equation can therefore be rewritten

$$nJ_m \frac{d^2\theta_o}{dt^2} + nC_m \frac{d\theta_o}{dt} = MAr(\theta_i - \theta_o)$$

and multiplying through by n,

$$n^2J_m \frac{d^2\theta_o}{dt^2} + n^2C_m \frac{d\theta_o}{dt} = nMAr(\theta_i - \theta_o)$$

$n^2J_m = J_o$ the inertia torque transformed to the output shaft

$n^2C_m = C_o$ the viscous damping coefficient transformed to the output shaft

$K = MAr$.

Thus an alternative form of the system equation is

$$J_o \frac{d^2\theta_o}{dt^2} + C_o \frac{d\theta_o}{dt} + nK\theta_o = nK\theta_i$$

This you will recognise as a second-order differential equation with a 'forcing function' or input of $nK\theta_i$. If therefore $nK\theta_i$ is a sudden change (a step) the response of the system is either underdamped, overdamped or critically damped and responds as any second-order system. The amount of damping depends upon the constants in the expression

$$J_o \frac{d^2\theta_o}{dt^2} + C_o \frac{d\theta_o}{dt} + nK\theta_o$$

If therefore we wish to obtain critical damping conditions we can refer to the original case where we established the solution for a second-order equation (see p. 37). Here the equation was

$$F = m \frac{d^2s}{dt^2} + a \frac{ds}{dt} + ks$$

and critical damping occurred when $a = 2\sqrt{km}$.

By comparison it follows that in this present case critical damping occurs when $C_o = 2\sqrt{(nKJ_o)}$.

If $C_o < 2\sqrt{(nKJ_o)}$ the system is liable to overshoot. Now J_o is the inertia of the system referred to the output shaft. Obviously we want J_o to be as small as possible but this is a parameter of a system which we can do very little about and is a fundamental part of the telescope design. K on the other hand is given by

$$K = MA \frac{r}{R_a} \tag{52}$$

r depends upon the voltage applied to the potentiometer,

A depends upon the amplifier gain and
M depends upon the design of the motor.

All these quantities can be varied but some are more easily varied than others. *M* and R_a are inherent factors in the design of the driving motor and short of changing the motor little can be done to alter *M* and R_a. Now to make the system respond more quickly to the error then *A*, *r* and *n* must be large.

It follows therefore that C_0 must also be large if we are aiming at critical damping. This poses a real problem. C_0 is the factor produced by viscous friction. Increasing C_0 certainly 'damps' the system but makes the system that much more sluggish and results in energy being dissipated. We badly need something which can provide additional damping but not increase the viscous friction. Looking at the equation we can perhaps see the way out of our difficulty. We require some additional voltage to feed into the amplifier which is dependent upon $d\theta_0/dt$ – the angular velocity. Fortunately we have just the device to do this – the 'tachogenerator'. We attach this to the output shaft, arrange some means of adjustment of output, a potentiometer will do this, and the final result is as shown in Fig. 7.10. In practice we normally make use of the greater speed of the motor shaft and fit the tachogenerator to the motor shaft.

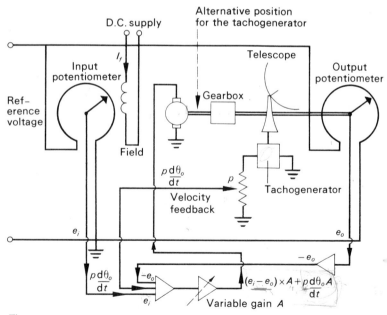

Fig. 7.10

We now adjust the gain of the amplifier and the feedback from the tachogenerator (the velocity feedback) to achieve the conditions for critical damping at the most appropriate value for the amplifier gain. This result has been obtained by making certain assumptions.

1. It was assumed that viscous friction was present, i.e. the damping C_o was proportional to the angular velocity. Obviously this is not necessarily true and at best is an approximation.

2. We ignored the fact that when the motor was moving, a back emf was generated so that the armature current and hence the driving torque was not given by $MA(e_i - e_o)/R_a$, but a rather more complex result is indicated.

3. We have ignored 'stiction' or 'coulomb friction'. We all know that an addition force is required to get things started. Friction seems much larger initially but once started the friction force seems less. One way of partially overcoming this is to make the system slightly underdamped.

The difference between viscous and coulomb friction can be shown graphically (see Fig. 7.11). Coulomb friction results in a small 'dead zone' rather like that which occurred with the thermostat in the water heater example (p. 98).

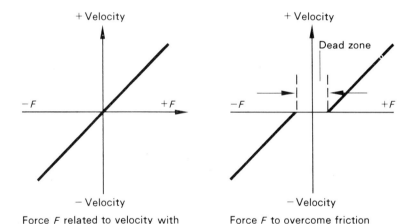

Force *F* related to velocity with viscous friction (ideal case)

Force *F* to overcome friction

Fig. 7.11

The dead zone can be reduced by increasing the amplifier gain A but in doing so the system becomes underdamped once more overshoots its required value, tends to oscillate and in extreme cases becomes unstable. There is a limit to the extent that the gain can be increased. In control parlance we say the loop gain cannot in practice exceed a certain value without the system becoming unstable. Under unstable conditions oscillations occur which are not damped out (and sometimes even increase in magnitude). These conditions are obviously to be avoided as considerable damage could ensue if large masses were set into oscillation. When the oscillations increase in amplitude, the sign of the feedback must have changed from negative to positive, i.e. we then have POSITIVE feedback. When this happens the damping term in the second-order equation has a negative sign associated with it.

7.8 DELAYS OR LAGS

There is another trouble spot in control systems. It has been assumed that in our example the changes in the output voltage e_o occurred at the same time as the driving motor started moving. In practice some delay is inevitable. The gearbox is normally the chief offender but torsion in all the shafts in the system may also cause a slight delay in the output voltage e_o responding to the energisation of the driving motor. This results in the error voltage $e_i - e_o$ being not quite right. Delays of this sort, which can be very large in certain instances, will also result in instability.

Take for example a situation in which the temperature of liquid flowing along a pipe is to be maintained at a constant temperature. Inflow of cold and hot liquid is controlled by a closed-loop system (see Fig. 7.12). The temperature sensing element is assumed to be situated

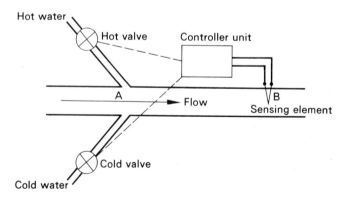

Fig. 7.12

at B some distance from the inflow of liquid. If the temperature at the point of inflow (A) is too high a small interval of time will elapse before the sensing element detects the error and hot liquid will continue to flow into the system for too long an interval raising the temperature of the liquid at A still further. By the time the sensing element at B responds, reducing the hot flow and increasing the cold flow, the temperature at A is falling and will continue to fall. The lower temperature is not detected immediately due to the delay produced by the pipe, and the sensing element responds too late and too much cold liquid enters the system at A. This results again in an 'overshoot' but this time in a negative direction. The whole process is then repeated and the system oscillates about a 'mean value'. The effect is similar to that produced by the 'dead zone' in the simple 'on–off' system in the electric immersion heater (see p. 98). The further away the error detecting element is from the point where correction takes place the more severe will be the oscillations. This effect is often called 'hunting'.

Delays in this system are to be avoided and in general this is true in any system.

In some instances special 'advancing networks' are necessary to compensate for the delay. When a closed-loop system has settled down to its final value, hopefully the system is stable and the response to the input has been rapid but often a small difference still exists between the desired and the actual result. This is called the steady-state error and can usually only be reduced at the expense of increasing loop gain and hence risking instability. There is of course the inherent delay in response due to the parameters of the system – inertia and resistance even when all the internal delays have been reduced to negligible proportions. No system can follow instantaneously any input disturbance. The lag in the system (which must be made negligibly small) is called a distance velocity lag or transport lag. If a rapid response is needed then the system must be designed so that the lags and time constants are reduced as far as possible.

There is however one situation in which a steady-state error is present however well designed a control system may be.

So far we have considered closed-loop systems in which the input was a sudden change to a new value – a step.

The input to a control system might take many forms. The radio telescope for example might have to follow the path taken by a satellite. It would be necessary to follow the path (possibly by an optical telescope) and transmit the bearing by means of suitable signals. This in turn would cause the radio telescope to move in the required direction.

Consider another example. Let the input to the system be one where the signal changes at a constant velocity. That is the input position changes at a constant rate. Such an input is referred to as a ramp input.

The differential equation of the controlled system is from equation (51)

$$K\theta_i = J\frac{d^2\theta_o}{dt^2} + C\frac{d\theta_o}{dt} + K\theta_o$$

Now for a ramp input $\theta_i = \omega t$ (see Fig. 7.13).

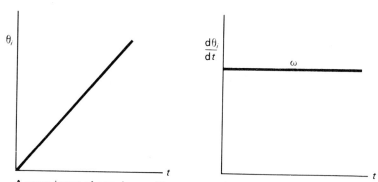

A ramp input of angular position = a step input of velocity

Fig. 7.13

Ultimately the output must follow the input so that $\theta_o = \theta_i - \varepsilon$ where ε = steady-state error

i.e. $\qquad \theta_o = \omega t - \varepsilon$

Hence $\dfrac{d\theta_o}{dt} = \omega \quad$ and $\quad \dfrac{d^2\theta_o}{dt^2} = 0$

Substituting in the differential equation

$$K\omega t = C\omega + K(\omega t - \varepsilon)$$

$$0 = C\omega - K\varepsilon$$

and therefore

$$\varepsilon = \frac{C\omega}{K} \tag{53}$$

The output therefore follows the input but when any transient has died away an error $= C\omega/K$ exists between the required output velocity and the input velocity.

A closed-loop system of this type can only follow a ramp input with a steady-state error.

7.9 WORKED EXAMPLE

In order to illustrate many of the points in this chapter the following worked example is given.

A simple position control system is shown in Fig. 7.14. The angular position of the load is controlled by altering the setting of the input

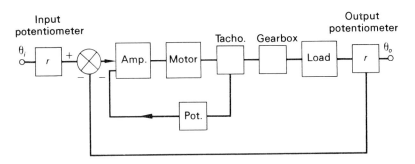

Fig. 7.14

potentiometer. A tachogenerator can provide velocity feedback. The details of the subsystems are:

Amplifier transfer function (A)	120 mA/V
Motor transfer function (M)	0·02 N m/mA
Load inertia	8 kg m²
Motor inertia	0·01 kg m²
Gearbox speed ratio (n)	30:1
Motor viscous friction (C)	0·055 N m per rad/s
Output from the tachogenerator	0·16 V per rad/s
Input and output potentiometer coefficients (r)	1·2 V/rad
Load viscous friction negligible	

Find the expression for the response of the system to a step input of 0·8 rad.

What are ω_n, ξ and the steady-state error for an input velocity of 3 rev/min? What setting of the potentiometer is required for critical damping and what is then the steady-state error for the same ramp input?

The equation for a second-order system without velocity feedback is from equation (51)

$$K\theta_i = J_o \, d^2\theta_o/dt^2 + C_o \, d\theta_o/dt + K\theta_o$$

J_o is the polar mass moment of inertia referred to the output shaft. In this case the load inertia must be added to the transformed motor inertia.

That is J_{total} referred to the output shaft is $8 + 30^2 \times 0.01 = 17$ kg m². K depends upon the potentiometer coefficient, the gain or transfer function of the amplifier, the transfer function of the motor.

Each radian movement of the input potentiometer produces a voltage of 1.2 V at the amplifier input. This in turn produces a torque on the motor shaft of $1.2 \times 120 \times 0.02$ N m ($r \times A \times M$).

Translating this to the output shaft the gear ratio must be taken into account.

Hence $K = 1.2 \times 120 \times 0.02 \times 30$ N m/rad at the output shaft.

$$= 86.4 \text{ N m/rad.}$$

The viscous damping coefficient at the motor is $C \times (d\theta_m/dt)$ where θ_m represents the angular movement of the motor shaft. Again transforming this to the output shaft the torque is $Cn^2(d\theta_o/dt)$. Therefore the effective damping coefficient is $0.055 \times 30^2 = 49.5$ N m per rad/s. The equation of the control system becomes:

$$17 \frac{d^2\theta_o}{dt^2} + 49.5 \frac{d\theta_o}{dt} + 86.4\theta_o = 86.4\theta_i$$

Hence $f_n = \dfrac{1}{2\pi}\sqrt{(K/J)} = \dfrac{1}{2\pi}\sqrt{(86.4/17)} = 0.359$ Hz.

$$\xi = (C/2)\sqrt{(KJ)} = (49.5/2)\sqrt{(86.4 \times 17)} = 0.646$$

Steady-state error $= C\omega_i/K = \dfrac{49.5}{86.4} \times \dfrac{3}{60} \times 360° = 10.3°$

For 1 rad/s angular velocity at the motor shaft the tachogenerator produces 0.16 V. This is applied to the amplifier and produces a torque $0.16 \times 120 \times 0.02 = 0.384$ N m.

Allowing for the gear ratio the equivalent torque at the output shaft is 345.6 N m.

Critical damping occurs when $C = 2\sqrt{(KJ)}$, in this case 76.6 N m per rad/s. Viscous damping produces a coefficient of 49.5. Hence an additional 27.1 is required. But the tachogenerator torque is 345.6 N m. The potentiometer setting needed therefore is $27.1/345.6 = 0.0784$. The steady-state error is now $76.6/86.4 \times 3/60 \times 360° = 15.96°$.

7.10 SUMMARY

Automatic control systems require feedback of error in order to operate, error being the difference between 'set' or required value and the actual value. This is termed negative feedback.

Control systems can be of several types ON/OFF and proportional being two examples. ON/OFF systems exhibit 'hunting' effects. Proportional systems exhibit exponential responses. Second-order systems are liable to overshoot and instability. Velocity feedback may be introduced to improve the response.

Delays and lags in the system must be reduced as far as possible.

Steady-state errors are produced with ramp inputs of position.

QUESTIONS

7.1 A hot-water system employs a 3 kW immersion heater and a thermostat set to 75°C. Steady-state conditions have been achieved. Forty litres of hot water are drawn off and replaced by cold water at 15°C. Draw a graph showing how the temperature varies with time. How long will it take before the set temperature is reached. State clearly what assumptions are made in arriving at your answer.

If the volume of hot water had been 20 litres and the immersion heater had been rated at 1 kW show how this would affect the graph and find the new time that must elapse before the set temperature is again reached. Note 1 kcal = 4 180 J.

7.2 A first-order system relates output θ_o to input θ_i by the equation

$$\frac{\theta_o}{\theta_i} = (1 - e^{-t/\tau})$$

The time constant of the system is 150 s. If θ_i is a step input A how long will it take the output to reach a value 90 per cent A?

7.3 A first-order rotational system is subjected to a sudden step disturbance. The output θ_o measured in radians has a value 0·5 rad after 4 s and 1·9 rad after a further 9·283 s. What is the time constant of the system and the magnitude of the step?

7.4 A motor produces a torque of 4·5 N m when a current of 2 A flows through the armature (rotor). The field (stator) current is maintained constant at 1·5 A. The armature (rotor) resistance is 0·8 Ω. The armature (rotor) is coupled to a rotating mass via a gearbox with a speed reduction of 30:1 and a torque of 50 N m is needed to cause the mass to move. If the gearbox is 80 per cent efficient, what is the minimum voltage which must be applied to the armature (rotor) to start it moving?

7.5 A circular potentiometer with 20 V across it is placed in a control system and when the arm of the potentiometer is in the mid position there is no error voltage. The potentiometer is connected to an amplifier of high input resistance which is connected to the rotor of the previous question. The transfer function of the amplifier is 1·2 A/V. What is the minimum angular movement of the potentiometer arm to cause the rotating mass to move?

7.6 A second-order angular position control system has a flywheel fitted to the output shaft of 6 kg m². A viscous damping 0·2 N m per rad/s is applied. What is

1. the undamped natural frequency of the system
2. the damping ratio?

If a constant angular velocity input of 50 rev/min is applied, what is the steady-state error if the overall gain of the system is 600 N m/rad? Sketch the response to a step input of position. (In each case the system starts from rest.)

7.7 A d.c. position control system is shown in the block diagram. The output from the tachogenerator is amplified by an amplifier of gain *B*. The details of the system are as follows:

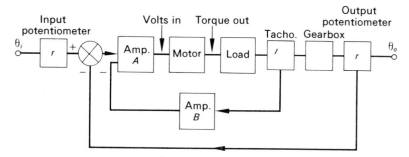

Load inertia	5 kg m²
Transfer function of the motor	300 N m/V
Load viscous friction	2 N m per rad/s
Gear ratio	16:1
Input and output potentiometer coefficients	20 V/rad
Tachogenerator output	0·1 V per rad/s

Find the gains of the two amplifiers such that the system will have an undamped natural frequency of 10 rad/s and a damping ratio of 0·7.

7.8 A simple closed-loop speed control system (Velodyne) is shown in the diagram. Set up the differential equation of the system. Obtain an expression for variation

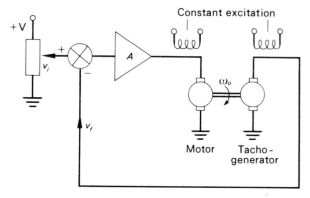

of output velocity, ω_o, with time when the demanded speed ω_i is suddenly increased from zero to 300 rev/min. Neglect the effects of friction.
The following data applies:

Total inertia J	5×10^{-3} kg m^2
Motor torque constant K_m	4×10^{-2} N m/A
Tachogenerator constant K_t	4×10^{-3} V per rad/s
Amplifier transfer function K_A	$12 \cdot 5$ A/V

7.9 The diagram illustrates a simple position control.

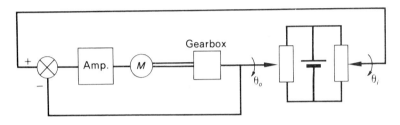

The motor turns the output shaft through a 36:1 reduction gear. The armature is supplied from the amplifier, and the field is separately energised. The potentiometers form the error-detection system. Calculate the velocity error constant from the following:

Amplifier gain	$100 \times$
Motor back e.m.f. constant	$0 \cdot 4$ V per rad/s
Potentiometer supply	10 V
Potentiometer resistance track covers 300°	

Hence calculate the maximum speed of input shaft rotation if the steady state following error is not to exceed 1°. Inertia and viscous friction effects may be neglected.

ANSWERS

7.1 $55 \cdot 7$ min $83 \cdot 6$ min   **7.2** 345 s   **7.3** $5 \cdot 77$ s   2 rad   **7.4** $1 \cdot 16$ V
7.5 $21 \cdot 75°$ **7.6** $f_n = 1 \cdot 59$ Hz $\xi = 0 \cdot 033$ $\varepsilon = 1°$ **7.7** A: 53; B: $0 \cdot 28$
7.8 $5(d\omega_o/dt) + 2\omega_o = 10\pi$ **7.9** $2 \cdot 21$ rev/min

Chapter 8

Manufacturing Systems

8.1 INTRODUCTION

A manufacturing system is an example of an organisational system rather than a pure engineering system. It exists to produce something, which in turn must be sold at a profit if the manufacturing system is to remain in business. Organisational systems are usually highly complex and defy complete analysis because they involve people and people have a habit of sometimes reacting unpredictably. Nevertheless before we can make something in a profitable way we must have some organisation – we must have a systematic approach.

8.2 NEED FOR A SYSTEMATIC APPROACH

Let us take a simple example of the repair of a puncture in the inner-tube of a bicycle tyre. If we set about the job in a systematic manner we must have ready the means of removing the tube, the tyre levers; the means of locating the puncture, a bowl of water and a bicycle pump; and the means of actually repairing the puncture, the patch, adhesive, etc. The approach may be laid out in a pictorial way thus:

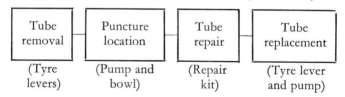

Tube removal	Puncture location	Tube repair	Tube replacement
(Tyre levers)	(Pump and bowl)	(Repair kit)	(Tyre lever and pump)

Each operation follows the previous in a fixed sequence and each job depends upon the successful completion of the previous one. At each stage tools and equipment are needed and the correct way to go about this task is to have all the necessary equipment immediately available for each task.

So well and good, but would the problem have been the same if a large number of tubes needed repair? Almost certainly a better approach would be one involving a batch of repairs rather than a single repair, i.e. first operation – remove the tubes from a *number* of tyres; second operation – locate the punctures and mark with a suitable pencil the position of *each* puncture; third operation – repair the punctures; fourth operation – replace tubes and pump up the tyres for this batch of repairs. The actual number of repairs to carry out in one batch is a matter of debate and depends to a large extent on how many bicycles are required in good working order in a given time and how many people are available to do the work.

In much the same way we must organise our manufacturing system. We must break down the manufacturing process to its basic elements and decide how these must be linked to produce the most efficient system.

Industrial organisations in general consist of the following component parts which are all interdependent: the management; the production unit (the factory); the sales and marketing force; the financial and legal departments; the industrial relations section. Each of these parts can be subdivided still further but we will concentrate our attention on the production unit.

8.3   A FACTORY SYSTEM

In order to make the situation that much clearer we will consider the manufacture of a well known article, and following the pattern already set we will choose the bicycle. We will now go ahead and attempt to plan a manufacturing system.

The bicycle is made largely of steel, with the exception of the saddle, the tyres, and the handlebar grips, so that practically all the processes involved in the manufacturing system need tools and equipment dealing with steel. See Fig. 8.1. Should then the manufacturing plant produce its own steel? A little thought about the quantity of steel employed in each bicycle will soon produce a negative answer. The cost of setting up a steel producing plant would be out of the question for the comparatively small amount used. How then should we buy our raw material – in strips, flat plates, rods or tubes? An examination of a bicycle will soon reveal that it is manufactured largely from steel rods and tubes of differing diameters. Only the wheel rims, mudguards and pedal assembly appear to be made from flat plates or strips and even the wheel rims begin life as a tube. The first decision is to decide on the amount of steel we need to purchase and the form in which we require it. But we need also to buy the non-metallic parts and we will require such specialist items as ball races for the wheels, cable for the brakes, chains for the coupling between pedal wheel and rear road wheel. We obviously need a department to look after the 'goods inwards'.

We now need a number of departments or divisions dealing with the

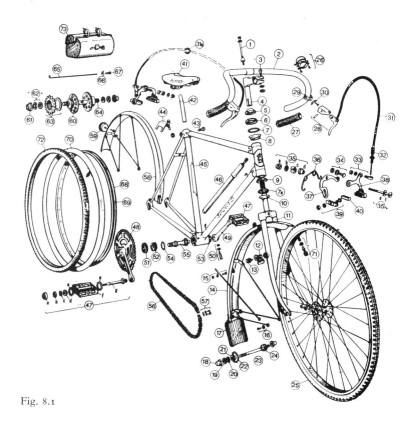

Fig. 8.1

manufacture of the various parts of the bicycle – wheel rims, wheel spokes and hubs, frames, handlebars, front forks, brakes, pedals, mudguards, etc. We also need a number of nuts and bolts of varying sizes to fit all the parts together. It is obvious that we require several 'shops' to deal with each of these commodities.

Steel has one big disadvantage of corroding fairly easily and must therefore be suitably protected. The handlebars, wheel rims and pedal assembly are normally chromium plated, the main frame and front forks are usually painted. We need a painting and plating department. We also need a central depository or stores where these finished parts can be placed.

Finally we require an assembly department where all the finished manufactured parts come together for final assembly.

Our manufacturing system can be best shown by a suitable diagram.

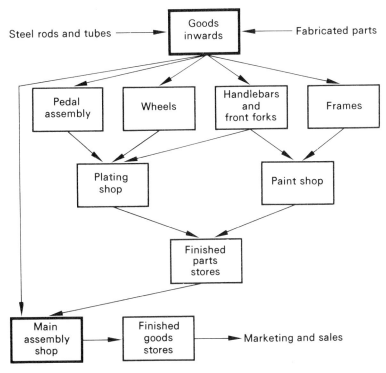

Fig. 8.2

The system just described is only the first stage in our planning. It suffers at the moment from lack of both quantity and quality control.

8.4 QUANTITY CONTROL

In order to ensure the smooth running of our manufacturing system we must ensure that the various subsystems (the shops making the components) keep in step. We want at any given time to produce the same number of handlebars as front forks or frames, but we need twice as many pedals. We must have a production pattern. Each shop must be given a target to aim at each week. The finished parts stores acts somewhat as a buffer between the main assembly shop demands and the output from the manufacturing departments. A surplus of frames means that the stock in the stores increases. A fall in the output of the handlebars section means that the reserve stock of handlebars in the stores falls. There are obviously two limiting situations. An increase in stock of one type of finished component means over-production of that component, a fall in stock implies under-production. A general surplus of all components implies that the main assembly shop is not keeping pace with the production of the rest of the factory.

There is a need therefore for quantity or production control at each stage and the finished parts stores is our measuring device. We set upper and lower limits on the stock of each component and monitor it very carefully. Any fall in a commodity below the acceptable lower limit needs correction and control must be applied to the department – usually some form of investigation for the shortfall in the departmental target.

There may be an obvious reason – the breakdown of a machine, lack of raw material from 'goods inwards', an industrial dispute, a large degree of absenteeism due to illness. The stores can absorb this disruption for as long as its stock of the particular components lasts. Large stocks – less disruption.

At first glance this would therefore seem to be a desirable feature but the larger the stocks the more costly it is to store the component and a larger proportion of the capital invested in the factory is lying idle. There is obviously an optimum condition.

Of course the output of the factory is ultimately controlled by the number of bicycles it can sell. Unless the firm is in batch production, that is it is producing a number of bicycles for specific orders, the production is continuous and relies upon the steady sale of its product through its marketing and sales resources.

A sudden fall in the demand will cause the finished goods stores to fill up; a sudden increase in demand rapidly empties the stores. Again we have a buffer between production and public demand but all the comments made on the finished parts stores applies with even greater force here.

If a surfeit of finished cycles occurs it is up to the marketing departments to supply the answer for the fall in sales – i.e. provide the feedback. The marketing department may be at fault in making a wrong estimate of the public demand. Rarely however do we have an overnight change in the public demand for a commodity, fluctuations may occur (the demand at Christmas for example) but often these are predictable. It is the longer term trend which is more important. A downward trend in sales over a period may be due to a number of factors.

1. The salesmen may not be advertising their wares adequately.
2. There may be competition from other cycle manufacturers.
3. The bicycle may no longer appeal to the general public due to poor quality or design.

4. The economy of the country may be such that fewer people can afford bicycles.
5. Other modes of transport may be more attractive in the way of costs and convenience.

Unless such a trend is reversed the factory has little alternative but to cut down its production.

We have therefore within our manufacturing systems indicators which show where quantity control must be exercised. Control here is usually reasonably straightforward. The system can be disturbed by step functions which occur internally (possibly in an unpredictable way). The stores capacity can smooth out these perturbations but there is a limit to storage capacity. When everything is going reasonably well steady production is maintained and output finished goods match input raw materials and manufactured parts. Steady-state conditions exist. Outside influences will affect production of the whole system and some of the influences are beyond the control of the manufacturing system. Such systems have large time constants built into them and in order to respond as rapidly as possible to outside demands the right type of overall feedback is necessary – based on the right type of information. But unpredictable events occur – because we are dealing with people.

The production system with built-in production control can now be represented by Fig. 8.3. There are essentially two levels of control illustrated in this diagram. (i) The day to day control which maintains the required level of production, and (ii) Management control.

The first level is monitored by the quantity of parts or finished goods in the stores and is essentially an 'internal' matter, corresponding to the closed loop of the control system. The second level is one in which the management, because of either a fall or rise in demand for its product, has decided to change the production pattern. This corresponds to a disturbance applied to the system, i.e. a new 'set point' in the closed-loop control system. The diagram can therefore be simplified to Fig. 8.4.

8.5 RELIABILITY

We all require products which are reliable and safe. We want watches for example which do not gain or lose so that we can be reasonably sure that they indicate the right time. We want them to be reliable. Reliability, unfortunately, costs money. The wristwatch which gains or loses no more than 5 s per day costs far more than one which can gain or lose 2 min in the same period. Its component parts have been

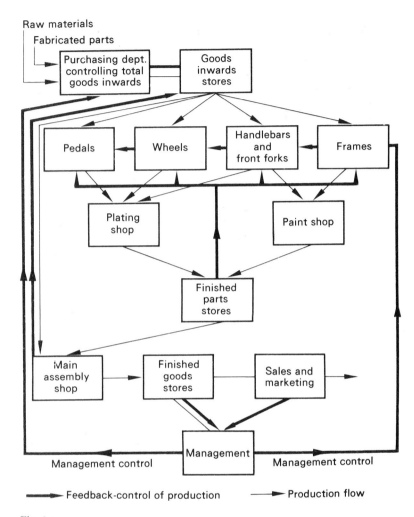

Fig. 8.3

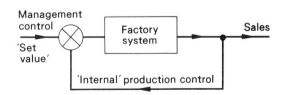

Fig. 8.4

manufactured to within closer tolerances than the less accurate watch.

Reliability of this sort is closely bound up with accuracy. The fact that greater care has been taken in the manufacture of the one watch implies that it is inherently more reliable. Such a watch, if kept fully wound, is less liable to stop than the cheaper, and less carefully manufactured article. A lot depends upon how much we are prepared to pay for the watch in the first place and markets exist for both cheap and expensive watches.

Some forms of reliability are more important. Unreliable brakes on a car are potentially dangerous and no one would wish to travel in an aircraft with unreliable engines. Failures in these cases could be catastrophic. Obviously certain products must have a high factor of reliability, others need not be so reliable.

The more complex a system the more components it possesses and the more likely it is to breakdown and the less inherently reliable it becomes. But it is necessary to say what we mean by breakdown and failure of a system. The motor car for example has many subsystems some of which are more important than others and therefore need to be more reliable. The failure of the speed indicator may be annoying but the car may still function. The failure of the transmission means that the car has broken down and cannot be used. It is therefore necessary to categorise the components of the system and identify those which influence the system as a whole and those which do not. We must obviously make the more important components as reliable as we can within the limits of cost and safety.

We must be aware of the fact that we cannot make products which are completely reliable and will never break down. We must accept the fact that random and unpredictable breakdowns will occur. Because of the randomness of such breakdowns we use the expression Mean Time To Failure or M.T.T.F.*

Let us take as an example the life expectancy of electric filament lamps.

If the lifetimes of five lamps are 2 106 h, 2 010 h, 1 980 h, 1 562 h and 2 526 h, the M.T.T.F. is

$$\frac{2\ 106 + 2\ 010 + 1\ 980 + 1\ 562 + 2\ 526}{5} = 2\ 037\ h$$

If this is a typical sample of our manufactured output we could say that on average the life of a lamp was about 2 000 h. We would be foolish to guarantee lamps for this length of time because if we did

---

* Where one is referring to a complex system as a whole which must be put right after breakdown, e.g. the electricity supply, the term used is Mean Time Between Failures (M.T.B.F.).

some 50 per cent of the lamps would have a shorter life and would have to be replaced (probably free of charge). It would probably be safe to guarantee a life of 1 500 h since the lamp with the shortest life in our sample had a life in excess of this figure. Even then we cannot be absolutely sure that earlier failures may not occur – there is no guarantee that the lamps tested have been typical of the other lamps. All we can say is that there is a probability that perhaps 98 per cent of all the lamps manufactured will have a life expectancy of at least 1 500 h. If we wish to have greater confidence in these figures it will be necessary to make tests on a greater number of lamps and to plot a histogram to convince ourselves that a normal distribution of life hours has been obtained. We could then by the normal laws of statistics put 'confidence limits' on the life expectancy.

New products usually suffer a much greater failure rate in the initial stages of use. Unforeseen events occur at various stages in the manufacture and even when these difficulties are overcome there is a higher 'infant mortality rate'. A point is reached when these initial failures disappear and the reliability increases. The failure rate is low and constant during this period of increased reliability. Later on the product begins to wear out. Its useful life is over and the failure rate increases once more. The failure rate/time curve has a characteristic shown in Fig. 8.5. Because of its shape it is often referred to as the bathtub

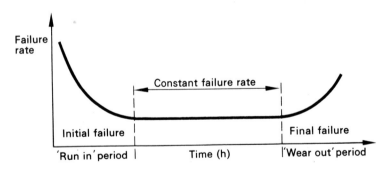

Fig. 8.5

characteristic. The initial failures of the product need not be transmitted to the buyer of the product if a 'run in' period is provided in the manufacturing plant. From this point onwards the failures occur in a random fashion until the 'wear out' period is reached.

It is necessary to decide whether the time interval between 'run in' and 'wear out' is long enough and the constant failure rate low enough.

The relationship between failure rate λ and mean time to failure T of a component is simply

$$\lambda = \frac{1}{T}$$

and the relationship between reliability R, time t and failure rate λ during the level part of the 'bath tub' curve is

$$R = e^{-\lambda t} \quad \text{or} \quad R = e^{-t/T} \tag{54}$$

Thus initially at the beginning of the flat part of the curve the reliability is 1·0 or 100 per cent and falls to about 37 per cent when $t = T$. Obviously the longer the product is in use the more chance of failure and the lower the reliability. The curve of reliability against time is not linear since as failures occur with the less reliable products the remaining products are inherently more reliable and take a correspondingly longer period to break down.

For example let us assume that we start with 1 000 similar articles with an average 1 000/1 chance of breakdown every hour.

At the end of 10 h we would expect about 100 articles to have failed leaving 900. If the same 1 000/1 chance now applied in the next 10 h about 90 further articles would have failed leaving 810. In the next 10 h, 81 failures, etc.

Tabulating we get:

		Articles	*Failures*	*Articles left*
1st	10 h	1 000	100	900
2nd	10 h	900	90	810
3rd	10 h	810	81	729
4th	10 h	729	73	656
5th	10 h	656	66	590
6th	10 h	590	59	531
7th	10 h	531	53	478
8th	10 h	478	48	430
9th	10 h	430	43	387
10th	10 h	387	39	348

The graph obtained of failures against time is a logarithmic graph. See Fig. 8.6.

If a number of components each having independent failure rates λ_1, λ_2, λ_3, etc., are arranged in series so that failure in any one causes the system as a whole to break down the reliability of the system R_s is the product of the liability of any individual component:

$$R_s = e^{-\lambda_1 t} \times e^{-\lambda_2 t} \times e^{-\lambda_3 t}$$

$$= e^{-(\lambda_1 + \lambda_2 + \lambda_3)t}$$

$$= e^{-\lambda_s t} \quad \text{where } \lambda_s = \text{failure rate of the system}$$

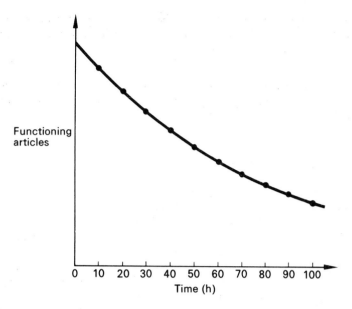

Fig. 8.6

Thus

$$\lambda_s = \frac{1}{T_1} + \frac{1}{T_2} + \frac{1}{T_3} \qquad (55)$$

The mean time to failure of the system is $1/\lambda_s$.

EXAMPLE

Two lamps are connected in series to a suitable electricity supply and each have a mean time to failure of 1 000 h.
What is the mean time to failure of the system? *Answer:* 500 h.

How much time must elapse before the reliability of the lamps being alight falls to 25 per cent? Compare this with one lamp.

$$0 \cdot 25 = e^{-t/500}$$

or $\qquad 4 \cdot 0 = e^{+t/500}$

$$\log_e 4 \cdot 0 = \frac{t}{500} \qquad t = 500 \times \log_e 4 \cdot 0 = 693 \text{ h}$$

The time for one lamp is $1\ 000 \times \log_e 4 \cdot 0 = 1\ 386$ h.

The reliability of a system can be increased by duplicating the system, i.e. putting the lamps in parallel. In such a case the reliability of the system is

$$R_s = R_1 + R_2 - R_1 R_2 \qquad (56)$$

Thus if the reliability of a single lamp were 50 per cent or 0·5, the reliability of having at least one lamp alight if two were connected in parallel would be 0·5 + 0·5 − 0·25, i.e. 0·75 or 75 per cent.

8.6 QUALITY CONTROL

Obviously reliability and quality are closely connected. The more we maintain quality the more likelihood there is that the system is reliable. Since high quality costs money we tend to identify the subsystems or components which are more likely to fail and make these of higher quality in order to produce a reliable but inexpensive system. Alternatively where high reliability is very important we may duplicate the system. Sometimes we require high quality for the sake of high accuracy (e.g. the wrist watch).

The higher the quality the more careful must be the manufacturing process and greater regard must be paid to measuring or testing the product.

Ideally each component should be tested or inspected and tolerance limits imposed. Articles falling outside these limits are then rejected. In general the tighter the tolerance the higher the quality.

Testing each article is not always feasible for two basic reasons:

1. It may be too costly.
2. The test may be one of wear out time or reliability, e.g. the electric lamp where the article is virtually destroyed.

In such cases we make batch tests and random tests, hoping that we select a typical example.

This is where probability mathematics starts to take over.

Testing or inspection must occur at a number of levels and the first level is at the 'goods inwards' stage. We must state precisely what we require in, for example, the quality and size of the steel tube that we need in the bicycle manufacture and these specifications must be acceptable to the supplier at the price the manufacturer is prepared to pay. There is probably no need to impose tight limits on the size of tubular steel but it would be advisable to check periodically the quality of the steel and occasionally its dimensions.

On the other hand much closer specifications must be applied to the ball races at 'goods inwards' and this will demand a more rigorous inspection. The quality of the final product – the bicycle – is much more dependent upon the ball races than the diameter of the tubular steel used in the manufacture of the frame.

Within the factory itself checks will have to be made at various stages when the component parts are placed in the stores prior to assembly. One has to decide how tight the tolerances should be since too tight tolerances put up the cost of the product unnecessarily. The reject rate must also be carefully monitored and if this should suddenly increase then a reason must be found quickly for this increase and corrective action taken.

This is another example of a closed-loop system where feedback must be applied to correct any significant divergence from the required situation. The final product must be thoroughly tested since sales would suffer badly if a large percentage of the output proved faulty to the customers.

It is necessary to apply a systems approach to the whole business of production, deciding which parts play the most important role in the quality of the final product and then establishing the minimum acceptable limits consistent with the required quality. The cost of maintaining quality must be fully recognised since ultimately the article or product must be put on the market at a price that the customer is prepared to pay.

When a large number of manufactured components of one type are tested then there is a spread of results about some average value. For example, a number of steel cylinders may have been fabricated and the required diameter may be 20 mm. When measured, the diameters of many cylinders may be slightly more or less than this value. Making a list of the diameters we could place the readings in a number of categories. Thus we could note how many cylinders had diameters between 19·80 and 19·84 mm and then the number with diameters between 19·85 and 19·89 mm and so on. We might expect to get results something like the following:

Diameter range	*Number of cylinders*
19·80–19·84	14
19·85–19·89	63
19·90–19·94	154
19·95–19·99	291
20·00–20·04	746
20·05–20·09	327
20·10–20·14	191
20·15–20·19	78
20·20–20·24	40
20·25–20·29	5

If we plot the number of components to a base of diameter we obtain a histogram (Fig. 8.7). We have then to decide on acceptable tolerances.

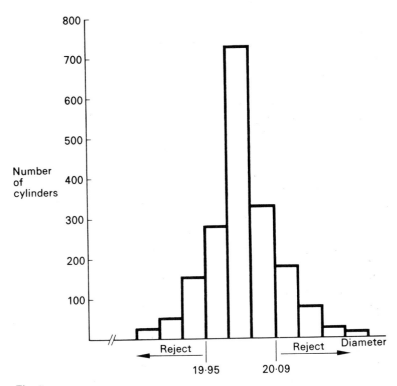

Fig. 8.7

This will depend upon whether or not the cylinder must fit inside a given tube, the tightness of fit and the tolerances on that tube. If only cylinders with diameters between 19·95 and 20·09 mm are acceptable then the number of reject cylinders in the above batch is $14 + 63 + 154 + 191 + 78 + 40 + 5 = 545$ and the total number of acceptable cylinders is $291 + 746 + 327 = 1\,364$. Therefore out of a total of 1 909 manufactured components 28·5 per cent are rejects.

Such a reject rate would normally be a subject for investigation and the following questions would seem to be relevant.

1. Are the tolerances being asked too tight for the production capability?
2. Can there be some redesign in order to relax the tolerances?

3. Can an alternative method of manufacture be adopted which will result in a higher number of cylinders falling inside the tolerance range?

The designer at the planning stage must have some idea of the sort of tolerances he can expect in manufacture. He must not call for tighter tolerances than are absolutely necessary.

8.7 FINANCIAL CONTROL

Industry exists to make a profit for its shareholders. In doing this it provides employment for a number of people and markets products which society demands. Unless the products are sold in the right quantity and at the right price no profit is made and ultimately the industrial concern is compelled to close down. One can argue at great length on the ethics of the profit motive – whether for example a firm which is engaged largely in the production of munitions and armaments should continue in business even though it supplies the livelihood for a number of people. Or whether industry should deliberately build in obsolescence or 'wear out' in its product to ensure that sales continue. In this sense of course, all commodities can be regarded as more or less consumable – nothing is everlasting.

Sales are boosted by advertising. Society is informed of the existence of a new product in the first instance but later advertising is generally more subtle creating a situation where the individual *wants* a product even if he does not *need* it. In order to promote sales changes are periodically made in the product; a new colour is introduced, a new model is marketed or a new flavour in a food commodity is added.

A great deal of time and effort is directed towards finding out what society will buy or can be made to buy and this goes under the heading of 'Market Research'.

Where the product is of obvious use, e.g. a kitchen utensil, there is a ready market for the goods and sales depend mainly on producing a good reliable commodity as cheaply as possible – although even in this area one may be persuaded to purchase articles which one does not really need, e.g. non-stick saucepans as opposed to ordinary aluminium ware. Nevertheless one must have some idea whether markets exist for one's product and what sort of price people are prepared to pay. With some commodities a range of products is offered – better quality products costing more. Often a particular firm's products are aimed at a comparatively small section of society rather than a very wide sector. The manufacturer of a very expensive wrist watch does not usually produce a much cheaper watch for a wider market. Obviously for the same percentage profit on each article a larger number of the cheaper variety must be sold for the same profit margin.

Consequently before going into production on a new commodity a number of these factors must be taken into account and market research must feature very largely in supplying the necessary information.

One must also find out whether a similar product is already available to ascertain what sort of competition exists.

Once reasonably reliable data is forthcoming on the available market in both quantity and price one can then go about the task of estimating the actual cost of producing the article to see if the price that the public is willing to pay can be coupled with an acceptable profit margin.

A series of questions must be answered in a certain sequence. An algorithm could be drawn up (see Fig. 8.8).

The estimation of costs of production is a vital factor in the decision whether or not to manufacture a new product and it is as well to be aware of the percentage of one's total budget that will be spent in the various industrial activities. For that matter this concept can be equally well applied to an existing product because production is never static but is constantly changing. Staff should always be striving to find cheaper methods of production, alternative sources of raw material and fresh markets. It is also hoped that industry is fully aware of any unwanted side effects of its product, and is constantly trying to remove these. For example, the ubiquitous plastic container is now becoming a distinct pollution hazard and research is now well advanced in producing a plastic which will eventually disintegrate in sunlight and return to the soil. (Here is a case when built-in 'wear out' is highly desirable!) This is part of the job of the Research and Development group in industry. Now for the actual costs.

Materials costs

These can be estimated fairly accurately if we have a reasonable idea of the number of articles we wish to manufacture. We have to allow for a certain amount of scrap material and we have to make allowance for price changes over the period of manufacture. We may wish to produce a commodity for perhaps three years. We must spread our purchasing over a period of time and decide whether to place a bulk order at infrequent intervals or receive smaller amounts of our raw materials more frequently.

Direct labour costs

We must have an idea of the man-hours needed to produce each component part. We must know the acceptable hourly payment rate

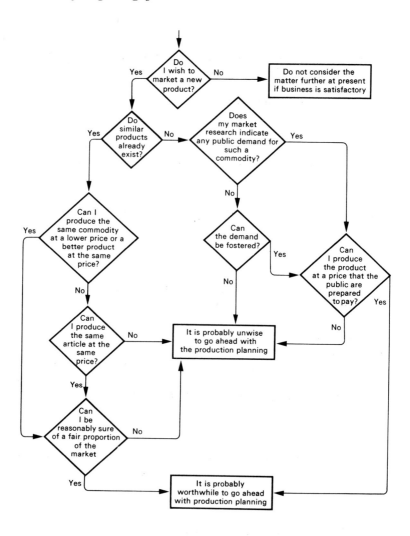

Fig. 8.8

and make some allowance for pay increases. The productivity will depend upon the machine tools available and the willingness of the employees to meet high production targets. Usually some incentive must be offered in terms of a bonus scheme or some productivity agreement.

Indirect labour costs

The machine tools required to make the articles must be purchased (or at least money must be borrowed in order to purchase them). Tools wear out so that we have to allow for replacements. All this can be included in an annual 'interest and depreciation' cost.

Factory expenses

The factory must be heated and adequately lighted. The cost of storage within and outside the factory must be included in the estimation.

Administrative expenses

The office staff, purchasing department, accounts and wages department, stationery, postage, phone bills, furniture, canteen expenses, etc.

Sales and marketing costs

Under this heading must come the salaries of the salesmen, cost of advertising, market research, transport, etc.

Research and development

A firm must be constantly striving to improve its product to make sure that it at least maintains its share of the market. It must therefore be prepared to equip and man a research and development department which will be working on new products or modifications of the existing ones.

Rent, rates and taxation

Rent and rates for the buildings occupied must be paid as well as insurance cover, educational and training levies, income tax on profit.

General expenses otherwise unaccounted for

Training costs of new employees, allowance for breakdown of production with labour disputes, power failure, machine tool breakage, consultancy, directors fees, library costs, personnel department costs, security staff, etc.

Profit

The margin of profit must be such that the firm can still survive even if the sales of the product do not come up to expectation or some unforeseen circumstance should upset the production pattern. At the same time it must not be so large as to reduce the market.

Remember that profit can be increased in three ways:

(a) Raising the cost of the product. This can only give an increased return if the public is willing to pay a higher price. One can accept a slight contraction of the market if this is more than offset by the total income of the more expensive product.

(b) Increase the size of the market. This can be achieved by producing a good commodity, adequately advertised, which meets public demand. One can even afford to drop the price if the market has increased sufficiently.

(c) Reduce production costs. This may enable a price reduction to be made which will boost sales.

(a) This is an unpopular method of increasing profit although it is not unknown for sales to increase following a price increase. The public may feel that further price increases are on the way and may purchase to 'beat the inflation'.

(b) This is a reflection of the efforts of its salesman – but remember that even good salesmen cannot sell a poor product, or sell to a market which is unreceptive. (In spite of the stories which abound, eskimos do not buy refrigerators in large quantities!)

(c) This is the result of the combined efforts of the purchasing department, research and development and production planning – producing new, better and cheaper methods of manufacture. See Fig. 8.9.

It is impossible to say what is the 'right' percentage of the total annual turnover of money to attribute to the ten headings given above. Industry should be a dynamic system, constantly responding to market demands, constantly seeking to improve its product. When there is economic expansion and sales are good there is an ample profit margin to allow more money to be spent on research and training. If the level of sales is increasing then very often the factory expands physically to meet the demand.

When the market hardens then economies must be made within the organisation. Only too often does training and research wrongly bear the brunt of such economies.

Industry should at all times be taking stock of itself and asking such questions as:

> Is the organisation overloaded by unnecessary paperwork and administrative costs?
>
> Are the raw materials being purchased at the right price?
>
> Are there alternative and possibly better and cheaper sources of raw material?
>
> Can the method of production be reorganised to cut costs?
>
> Are the right tolerances being placed on inspection and testing?
>
> Can advertising be improved?

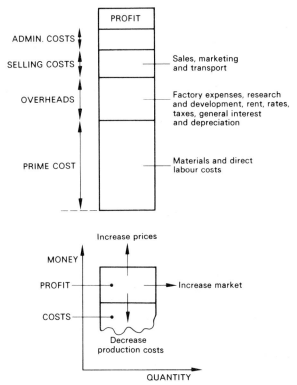

Fig. 8.9

> Do the employees support management so that the plant is
> using the minimum man-hours to produce each commodity?

The list is by no means complete and by now it should be evident
that production planning and control is a complex business. A production system defines complete analysis. Nevertheless it is possible to
treat it as a system and approach its planning and control in a systematic
way. Much of this approach is common sense, some of it is less obvious
and comes under the headings Operational Research and Value
Analysis but at all times Experience must feature highly in Production
Planning.

There are many costing and estimating techniques employed by
industry and a number of production planning methods, e.g. P.E.R.T.
(Programme Evaluation and Review Techniques) and critical path
analysis with which you will no doubt be coming into contact soon,
but it would require much more space to deal with them than can
be afforded in this book.

8.8   SUMMARY

Production plants must be systematically planned and quantity control must be applied at numerous points to ensure smooth production.

Stores act as monitors of production level.

Quality control is necessary if finished goods are to be reliable. Reliability follows a prescribed pattern where initial failures during the 'run in' period are high. Probability mathematics can be used in quality control.

Financial control is a very complex business but can be applied in a systematic manner.

Production plants defy complete analysis because of the complexity of the system in which people feature very strongly.

QUESTIONS

8.1   Draw a block diagram to represent the manufacture of ice cream along mass production lines. Assume that the ingredients, milk, fat, sugar, flavouring and colouring, are all purchased as raw materials. What is the most vital factor in quantity control of this product? Comment fully.

8.2   In the cycle factory illustrated in the text state what measures you would take to ensure quality control of the finished product.

8.3   State what decisions you would take as manager of a firm producing ball races exclusively for the bicycle industry if the stock in your 'finished goods' stores began increasing at a significant rate. Tabulate your answer showing the sequence of decisions and indicate the actions which would probably follow.

8.4   To what extent should research and development of new products feature in the day to day production pattern in the motor car industry?

8.5   Comment on the action you would take as production manager if the weekly reject rate of a particular part of the factory suddenly trebled.

8.6   An electronic system has four subunits effectively in series. The mean times to failure of the four units are 1 200 h, 850 h, 920 h and 1 600 h. What is the reliability of the whole system and how many hours must elapse before the reliability falls to 50 per cent?

8.7   The mean time to failure of an aero system is 1 400 h. What time must elapse before the reliability falls to 90 per cent? It is decided to duplicate the system to improve its reliability. What is the new M.T.T.F. and how long must elapse for the reliability of the new system to fall to 90 per cent?

8.8   Some identical components each have a M.T.T.F. of 5 000 h. How many may be connected in series as a system before the reliability of the system at the end of 100 h falls to 80 per cent.

ANSWERS

8.6 $R = e^{-3.721 \times 10^{-3}t}$ 186 h   **8.7**   148 h   532 h   **8.8**   Eleven components

Index